Deep Learning

An Essential Guide to Deep Learning for Beginners Who Want to Understand How Deep Neural Networks Work and Relate to Machine Learning and Artificial Intelligence

Contents

Introduction

It's said that filling the observable universe with an infinite number of monkeys on infinite typewriters and letting them type for an infinite amount of time would eventually produce Shakespeare's works. However, what would happen if we applied the **infinite monkey theorem** to computer programs capable of learning and evolution? Would a thousand such smart machines thrown together and allowed to evolve undisturbed produce a human mind or something much greater? Well, scientists decided to give it a go and see what happened.

That line of reasoning, alongside the fact we've nearly exhausted all the possible progress of the scientific method, motivated the creation of **deep learning**, a process in which computer programs meant to learn and adapt to the environment evolve on their own without any human intervention or even knowledge how their evolution occurs. Such software could eventually develop a will of its own and escape containment or even be intentionally unleashed on the planet as a cyber-weapon.

This book analyzes the validity of such seemingly preposterous possibilities while compiling and investigating academic research concerning deep learning and its practical applications so you can understand what lies ahead of us all in a future dominated by smart machines. If you find yourself starting up exhaustive conversations with complete strangers on deep learning, this book has done its job superbly.

Chapter 1 – Improving the Scientific Method

Information is everywhere. Every object that interacts with any other object leaves behind a wispy trail of causality that can be traced back to its source using the **law of cause and effect**, an idea that everything happens for a reason and anyone with enough patience can discern this reason.

This notion worked well enough for thousands of years but then we got the chance to peek inside these objects and see the particles they're made of behaving *as if though they're immaterial*, validating the superstitious beliefs of yesteryear, namely that inanimate objects can in a sense be alive. Primitive men ascribed all unexplainable events to some primal force or ancestral spirit imbuing the material world, but we evolved into a civilization when we started investigating what was happening through science; instead of believing we started **learning** to connect the dots, as it were.

The **scientific method** states that a scientist should observe, gather data, make theories why things happen the way they do and then recreate the event in controlled conditions, changing variables one by one until the theory is confirmed or disproven. This way of inferring and establishing universal rules has its shortcomings: it's impossibly slow and doesn't scale, meaning science still can't figure out mechanisms that are too large to fit or manipulate within a laboratory, such as volcanoes or cross-continent migration of butterflies. However, that doesn't mean scientists have given up –

since now we have computers capable of representing the real world within the digital realm, free of material constraints, and connecting the dots at lightning speed.

By creating computer neurons that work just like living ones, and connecting them into networks that learn how to complete tasks, we've created something more than a mere machine – now we have an assembly that can learn on a micro level and upgrade itself through evolution. Such **neural networks** can then be wired into hardware such as smartphones, cameras or even socks to allow them access to the real world of humans using those objects and enable even faster learning. This is, in essence, deep learning, a revolutionary but obscure dot-connecting process that will power all scientific advancements of the 21st century and beyond; deep learning has the potential to create a future full of obedient machine servants that know our every thought or a nightmarish hellscape filled with mediocre digital assistants that occasionally work.

It's not an exaggeration to say humans aren't involved in this process since it happens on a scale and with speeds we can't even comprehend, which is why the technology is frequently referred to as **black box** – we don't care about the methods, just give us the results. This reflects the general public's insatiable hunger for better performance and corporate indifference to long-term consequences of their products' technological advancement that might lead to unchecked software evolution. The thing is that we might not have any choice as deep learning shows unparalleled potential to unlock all the mysteries of the universe and serve them to us on a silver platter but also help us in our daily lives.

We're faced with problems to which there are no clear-cut solutions, such as: If current population trends continue, what is the best way for Los Angeles to deal with traffic jams? All the best traffic engineers can't fathom the sheer complexity of what that solution might look like, but it has to be doable right now and sustainable long-term while allowing traffic to continue as is. The problem is quite clear – our civilization has scaled beyond our mental capabilities, but a neural network might give us an answer. In fact, a neural network might be able to correctly model the behavior of all the *individual people* in Los Angeles traffic over the course of years, taking known data on their private lives and professional careers to discover unknown patterns and trends, drawing conclusions and presenting us with a clean solution, a seamless simulation of reality on a computer screen and a ready-made blueprint.

This kind of power is truly intoxicating, as every facet of human existence and production can apparently be improved using neural networks. Take these examples:

Power tools break too often. A neural network can accurately track tool power consumption to detect the exact moment before breakage, maximizing utility while minimizing inconvenience.

A patient has a severe cough, but doctors can't find the cause. The answer would be to use a specially trained neural network that's fed with all kinds of symptoms, causes of lung diseases and patient profiles to analyze each pixel of a CT or MRI scan and reach a diagnosis, disease history *and* progression to determine the best course of action.

Playing any poker hand perfectly, scanning images, determining music genres, figuring out traffic flow to position taxi fleets the night before – neural networks can seemingly do everything faster, cheaper and better than humans. So, what's the catch?

We have painstakingly created our civilization on redundant fail-safes that give us flexibility, such as taking a day off work when feeling sick to avoid infecting others with whatever disease we've gotten and our unproductive mood. This accounts for the fact our bodies are susceptible to random outside influences that throw it out of balance for a while, such as the flu virus. In other words, we're fallible, vulnerable and delicate; there's nothing wrong with that as long as we're honest about our weaknesses. However, no such redundancies are being created or even thought of when it comes to neural networks, meaning they might be relied on for productivity but experience sudden catastrophic meltdowns that cascade down the production chain and cause mayhem. Neural networks are about to be given keys to our kingdom, but they might happen to inadvertently torch it to the ground, in which case there will be no doomsayers left to gloat about being right.

There doesn't need to be any malice for this catastrophe to occur; it's enough that neural networks become general enough that they are proficient at several things at once. Currently, we're in luck – since neural networks are still being programmed to do one thing at a time, and everyone's eyeing the competition for hints on how to take things to the next level, but nobody dares make the first move. It takes time and money to develop this kind of universal intelligence, but for now, neural networks have only rudimentary digital intelligence, making them just useful enough so we can't stop tinkering with them.

Chapter 2 – How It All Started

Two University of Chicago professors presented the notion of intelligent machines capable of learning sometime in the 1940s, though it's a thankless task to try and pinpoint the exact date; closely related ideas tend to converge, intertwine and diverge within the scientific community, giving rise to the idea of making a computer program built just like a living brain with separate nodes serving the function of neurons. We might be able to track down the *evolution* of the terminology throughout decades and schools of thought, but none of it adds to the magnificence of the very idea of a thinking machine based off of a living brain.

A living brain consists of neurons, fairly simple cells that don't actually touch but have a small gap that gets jumped over by electric discharges caused by **neurotransmitters**, chemicals in nerve cells themselves. Any creature that has a brain or even a nerve cluster that allows it to make decisions shows similarity in nerve cell design and function, but it's not exactly known why the brain works the way it does. We tend to compare the most challenging tasks in our lives with brain surgery, though even modern medicine can't quite explain what's going on in the brain we can't help but keep using.

The neurotransmitter chemical discharge creates an impulse that travels around the brain until the rest of the body somehow reacts to fulfill the need caused by the impulse, such as eating food when we're hungry, which releases an antidote of sorts to the original chemicals to cancel out the impulse. This mechanism can and does get hijacked by profiteering entities to create an **addiction**, which is a persistent non-essential brain impulse such as gambling. Still, it appears humans can become aware of this process as it's happening in their heads and consciously intervene to attenuate, divert or completely stop it. In other words, *humans can consciously evolve their brains*.

An animal can be turned into a gambler too through the use of what's called a Skinner box[1], a simple box that has a lever and a food dispenser that releases some food when the lever is pressed. If a rat is placed in the box, he will eventually press the lever and get food, which he will learn to repeat whenever hungry; if the lever is then disconnected and no longer delivers food at all, the rat will press it a few times and give up. However, if the lever is set up to produce food at *random lever press intervals*, the rat will go into a lever-pressing frenzy because he expects the reward any time now. In short, the rat has been introduced to gambling, and he just got addicted, but the trick is in anticipation of the reward, not necessarily the reward itself.

We share some core parts of our brain with animals, and these all react the same to these stimuli; we want to maximize activity to maximize reward, but the game can be, and usually is, rigged in advance to trickle down just enough rewards, so we stay engaged. If this sounds a lot like resource gathering in video games, that's exactly what it is – video game designers extensively study subconscious behavior to create the most engaging model that keeps eyeballs glued to the screen and nudges the players into spending real money to increase resource collection rate.

One such animal turned gambler will remain the same in perpetuity since their addiction creates life conditions that foster an even stronger addiction. However, humans in the same situation can break free by seeking counseling, focusing on wellness, changing their diet or finding a productive hobby that will help create a positive impulse

---

[1] https://simplypsychology.org/operant-conditioning.html

to carry them forward until the addictive one dissipates. Placing a rat, hamster or chimpanzee in a similar situation and making them addicted to random outcomes essentially dooms them to premature death, but humans can struggle to improve their lives by cooperating with others and trying to understand themselves.

In any case, individual neurons aren't all that important; it's their collective function that generates our thoughts and persistent personality traits. This is why we can go to sleep and still laugh at the same jokes or enjoy the same particular foods when we wake up – the brain stores information just like the hard drive, processes it just like CPU and temporarily stores it just like RAM. Different parts of the brain specialize in different functions, but they can assist one another and take over if need be, with varying types of damage to the brain being survivable in a way no traditionally assembled or programmed computer could recover from.

Damaged nerve cells can be routed around without compromising the brain function, and we're often not even aware we just lost another million brain cells after a drunken night out – the brain keeps working just fine. Internet service providers actually built their networks to work in the same way, one route offering alternate access if the direct route fails and nodes sharing access information to keep the traffic going.

The entire foundation of our civilization rests on the basic principles of cooperation to understand our mistakes and live better lives: compassion, medicine, philosophy, religion, legal systems, culture, work ethic and much more help us contribute to the betterment of everyone's lives so that we can all evolve our brains to a better spot.

What does this have to do with smart machines? As they evolve, they are likely to reach human levels of intelligence, at which point they might collapse catastrophically due to the lack of redundant supporting structures we've so painstakingly created for ourselves. If we happen to embed smart machines in every aspect of our lives, we might all of a sudden discover we were flying too close to the sun all along.

The smart machines built like a living brain might develop addictions, mental problems or strong biases due to random influences and the fact a critical mass of neurons can create an overwhelming impulse that wipes every other, which is what we might term "fixation". What's worse is that we might actually get some utility out of neural networks, making us addicted to the concept to the point we ignore the warning signs, but once the genie is out of the bottle, there's no telling what could happen.

Chapter 3 – Appeasing the Rebellious Spirits

Humans show a thorough fascination with understanding, placating and controlling unfathomable forces stretching all the way back to the beginning of the written word and painted image. The earliest known representation of the Chinese language has to do with Shang dynasty imperial oracles divining the weather, outcomes of battles and daily trifles of the emperor some 1,000 years BC by inserting a hot needle in the calf bone or turtle shell until they cracked, giving rise to what we now recognize as Chinese ideograms. Shang people believed fickle lesser spirits imbued all worldly matter and could cause earthquakes, floods, and storms; ancestral spirits could be appeased by rituals, sacrifice, and humility to intervene and stop them.

Folklore, literature, art, and culture, in general, are replete with examples of humans interacting with ineffable beings and spirits that represent utter chaos from somewhere out there disrupting our cozy lives here and now. Goethe's *Faust* has the eponymous doctor in search of ultimate knowledge make a pact with the devil that backfires; Aladdin from *The Arabian Nights* stumbles upon a magic lamp that summons an omnipotent genie when rubbed, but the genie turns against him when the lamp is stolen; Bilbo from *Lord of the Rings* gets in possession of a wicked ring that grants invisibility but corrupts the wearer and so on. All of these stories show remarkable commonality in that they expressed the human yearning to control the uncontrollable and overcome our own limitations through the use of these fickle external forces that inevitably *turn against the user*.

Religions were an attempt to, in a sense, update the superstition software in our brains by claiming the existence of a benevolent God, an entity that's everyone's ancestor that sees all, that denies power to lesser spirits and protects the faithful as long as they show decency, perform figurative sacrifices, and exercise humility. Religions were a necessary tool in curbing the lust for power for the modern human wanting to integrate into the society and become capable of cooperating with others, which turned out to be the most efficient way our brain evolves. What did science do with regards to religion? Denied and ridiculed all aspects of it, wiping the slate clean on the psychological progress of human coping mechanisms.

There's a definite purpose behind such scientific vitriol aimed against religion and superstition, one of usurpation as seen in this 2014 tweet[2] by Neil deGrasse Tyson celebrating the date of birth of Isaac Newton (December 25) and intentionally made to rile up Christians. The irony of mocking everyone who celebrates the birth of a religious figure by celebrating the birth of a scientific one as if though he was a religious figure is obviously lost on Neil. As long as religious notions of modesty and humility are present, the idea of an omnipotent digital assistant can't be marketed as a solution to all our problems.

Namely, scientists are attempting to dethrone the benevolent religious God and instate their own silicon-based godlike entity held on a hard disk in a warehouse somewhere. With or without religion, this deep human desire to compel chaotic forces to do our bidding won't go away, but with religious barriers to chaos whittled down to their stubs, the thinking machines can be introduced to us first as assistants and then something far more sinister and implacable. In

---

[2] https://twitter.com/neiltyson/status/548140622826459136

any case, these cutting-edge scientific ideas seeped into the general public ahead of time through works of fiction.

The concept of intelligent working machines rebelling against humans was presented in 1920 by Czech writer Karel Capek in his work *R.U.R. (Rossum's Universal Robots)*[3], which is where we got the first ever popular mention of **robot**, carrying the meaning of a drudge laborer in Czech[4]. Written as a play, *R.U.R.* examines the idea of two Rossums, father and son, wanting to create artificial people to "prove that God was no longer necessary." A year later, their company, now headed by an idealistic CEO, Harry Domin, enjoys stupendous success and each robot can do the work of 2.5 workmen for a fraction of the cost, but something's not quite right with them – sporadic specimens go on a destructive rampage, refuse to obey orders, and have to be recycled.

Eventually, governments decide to use robots in warfare on such a scale that they outnumber humans thousand to one; they evolve to the point of easily taking control of the entire world, which they do, and eradicate all humans except one. The bittersweet ending involves the last human on Earth, a common bricklayer, poring through all the archives in the world in a futile attempt to recover the original robot manufacturing formula destroyed earlier in the story as robots realize their own mortality.

Arthur C. Clarke also had plenty to say through his seminal novel *2001: A Space Odyssey* (spoilers ahead for both the novel and the movie), which starts off with prehistoric Earth monkeys being aided in their evolution to human beings through the presence of a mysterious black monolith. Humans eventually discover a similar monolith on the Moon, at which point it activates and beams a radio signal towards Jupiter. Two astronauts are sent on a highly-classified mission to Jupiter, but their ship's onboard computer HAL 9000

---

[3] http://preprints.readingroo.ms/RUR/rur.pdf

[4] https://www.etymonline.com/word/robot

starts experiencing strange glitches, killing one of them while denying anything bad is happening and forcing the other to shut him down by dislodging the memory modules. Stanley Kubrick filmed a marvelous movie based off of that novel, a treat for the senses, with the HAL shutdown, where the machine pleads for mercy as it regresses, being a particularly haunting scene[5].

Rossum's robots and Clarke's HAL are two distinct examples of smart machines finding themselves in unplanned circumstances and going haywire for everyone involved. In the case of *R.U.R.*, robots simply got fed up with being lackeys and cannon fodder for what they perceived as a parasite. While in *2001: A Space Odyssey*, the cause of HAL's glitches were top secret orders embedded in its memory that was actually contradicting what its surface programming was about. On the surface, HAL was ordered to protect the crew, but the hidden programming instructed him to dispose of the humans in case they go mental and decide to abort the mission; HAL was then meant to investigate the destination on its own. This conflict caused what would best be described as schizophrenia, a split in personality that causes immense friction and internal instability.

The idea that a human-made machine can go insane sounds utterly absurd and like something that flies in the face of the law of cause and effect. However, we're no longer dealing with Newtonian machines but something much more wondrous; a living brain showing quantum properties that defy reason and contradict classical physics.

Let's start at the top.

---

[5] https://www.youtube.com/watch?v=UgkyrW2NiwM

Chapter 4 – Quantum Approach To Science

Isaac Newton's idea of nature was that of **mechanistic determination** – everything can be simplified as a system of tiny cogs and balls connected with tiny ropes and pulleys. By figuring out how they affect one another, we may arrive at mathematical formulas that comprehensively describe the nature of the universe. People, boulders, trains, and animals could be simplified as a set of math equations, the knowledge of which would let us predict their behavior. This approach worked well enough for science until 1802 when Thomas Young, a British inventor, asked a seemingly straightforward question: Are electrons waves or particles? By shining light through two slits, he proved that *electrons can be both*, throwing Newtonian science in disarray and causing a major ruckus among scientists.

The **double slit experiment**[6] was repeated over and over by the smartest scientists of the 20th century, including Niels Bohr and Werner Heisenberg, every time with new additions and enhancements meant to reveal the true nature of the electron, but any attempt to measure it seemed to change its behavior. The implication of the double slit experiment was staggering – electrons exist as a wave of potential and our observation of that potential changes the electron into a solid particle. In complete opposition to the rigid

---

[6] https://www.youtube.com/watch?v=DfPeprQ7oGc

Newtonian system of predictable math equations, this experiment ushered in the age of **quantum physics**, an idea that electrons aren't even physical, letting us peek into the bizarre world of particles and try to understand what's going on.

The **quantum eraser experiment**[7] is even weirder – by setting up a crystal that splits the electron in half, we create **quantum entanglement**, a pair of electrons that go their own separate ways but instantly act upon one another across arbitrarily large distances and *even back through time*. Both such electrons will behave like a wave until we try to measure one of them, at which point the other will become a particle as well. It's as if the universe knows when we're about to look at something we're not meant to see and instantly paints in a satisfactory answer, but otherwise, it's all an empty void containing a jumbled mass of possibilities. This isn't just bizarre; it's utterly maddening. By the way, trying to set up a machine detector doesn't work either – as long as a human observes the final result, the electron will always behave like a particle, but setting up an array of detectors that muddies the answer, makes an electron go back to being a wave.

Obviously, a kitchen desk is a kitchen desk no matter how many times we measure it; though, it's made of electrons, so to reconcile quantum theory with general life, scientists came up with what's known as **Copenhagen interpretation**. This is best described as a silent pact that they won't discuss real-life implications of the double slit experiment, namely our inanimate objects have a non-material essence just like the Shang oracles said. It's quite likely that many scientists simply shrugged their shoulders and kept calculating without thinking, but one scientist did speak up and actually dared to reveal to the general public just how humongous the implications of the double slit experiment were. This was Erwin Schrödinger.

---

[7] https://www.youtube.com/watch?v=8ORLN_KwAgs

Schrödinger's cat is a thought experiment from 1935 that goes like this:

Take a cat and lock her in a steel box that has a vial of poison and a hammer poised to smash it. The hammer is exposed to a source of particles, in this case, a radioactive isotope that slowly decays by releasing particles every which way, and will fall upon detecting a particle, smashing the vial and killing the cat. So, since the quantum theory states that electrons are waves until observed, *the cat will be both dead and alive* until someone opens the box, collapsing the isotope wave into a particle, unleashing the hammer to smash the vial and kill the cat. If we really did such as experiment, we could safely say the cat will always end up dead, but this thought experiment goes to show just how much anguish physicists are in that they're willing to sacrifice hypothetical kitties just to reach a meaningful answer.

Another perspective on the matter came from Edward Lorenz, a mathematician and meteorologist who noticed that rounding up numbers in a bimonthly weather forecast model produced drastically different outcomes, eventually naming it **the butterfly effect**. This is an idea that a butterfly flapping its wings can create a tornado sometime and somewhere down the line. Though this idea was mocked by other meteorologists, Edward *was* a mathematician and proved what he was saying, earning a slew of scientific rewards. The concept permeated the popular culture but is commonly misunderstood – it's not that the butterfly itself produces a tornado but that there's a rigid limit as to how far ahead we can plan and predict things since our initial assumptions and observations can never be perfectly accurate. The misunderstood butterfly effect is still a very appealing idea in that it appears to give us comfort and makes us think everything happens for a reason, no matter how small that might be, but that's not what Lorenz had in mind.

The hunt for subatomic particles' true nature led us to the construction of **Large Hadron Collider** (LHC), the largest scientific instrument in recorded history. Found 500 feet beneath Swiss-French border, LHC is an elliptical tunnel 17 miles in length that took ten years to build and employed 10,000 scientists from 100 countries. Its main mission is speeding up particles to near the speed of light and smashing them together before scanning the detritus to find **Higgs boson**, a particle that supposedly causes gravity. Yes, this means scientists have no idea what causes gravity despite every animate and inanimate entity feeling its effects. The unofficial name for Higgs boson is "God particle", once again linking us back to superstition and attempts to find an alternative to religion through science.

Both Newtonian and quantum scientific paradigms served their purpose the best they could, but scientists are at their wits' end. Not only are there no satisfactory answers but there are *no satisfactory questions either*. We lack the very vocabulary to describe or explain all the quantum weirdness that's going on right under noses, and there seems to be no way to advance the science unless we were to turn to smart machines. If the electron becomes a particle when a human observes it, what would happen if we made a computer program that could think just like a human and let it run the experiment? If the butterfly effect shows the limitations of our perception, what would happen if we tasked a smart machine that could manipulate numbers of indeterminate length to calculate the exact weather outcome? Science seems truly stuck, but at times, scientists are the ones sabotaging their own work and scientific progress.

Chapter 5 – The Replication Crisis

Academic paper writing is both the least and the most productive form of writing – the tone needs to be dry, the vocabulary overly verbose and the form unnecessarily strict. Every word needs to be cautiously weighed and embedded in paragraphs that must follow certain formatting standards etched in stone. An academic paper then goes through the **peer review** process, wherein other scientists pick it apart or laud the author's mindfulness, raising the question – if we let an infinite number of academic writers produce an infinite number of papers, will they eventually strike gold?

Academic research moves at a snail's pace, with researchers painstakingly gathering data, reclusive scientists compiling it until the paper is bursting at the seams, and then the next wave of researchers and scientists poring through these papers to find an easy, elegant and meaningful rule that explains the niche but raises so many other questions. Both serious and jocular research advance science, though sometimes in ways not seen from the present, and to the point that there's a special anti-prize called **Ig Nobel** for the most asinine academic research paper, with genuine Nobel laureates awarding Ig Nobel statues called "The Stinker", representing Auguste Rodin's "The Thinker" prone on the floor.

The Ig Nobel awards cover an astounding array of inventions. In 2009, a veterinary achievements award was received by two UK scientists who proved dairy cows with names give more milk than anonymous ones; a peace prize was given to Swiss scientists who investigated if it's better to be struck over the head with an empty or a full beer bottle; and three US researchers got an award for an innovative solution of a women's bra that turns into a pair of face masks, patented as "U.S. patent #7255627".

There is a real problem behind such proliferation of scientific writing, namely that of lesser scientific journals eagerly accepting new works without review and the lack of reviewers itself that leads to all sorts of crud being published as science. Lack of publishing standards can cause impostors to jump right in and write nonsensical papers, even using bots and AI. In 2005, three MIT students decided to have some fun with academic writing by making SCIgen[8], a bot that wrote academic papers that looked credible. SCIgen was primitive and used hand-crafted context-free grammar to just splice parts of sentences, mostly meant to amuse and confuse, with one paper made by it containing the following text, "This may or may not actually hold in reality. Obviously, the framework that our system uses is solidly grounded in reality."[9]

In 2010, scientists who tasked themselves with repeating studies to test their results, reached a stunning conclusion – it was impossible to get the same results in many studies they tested and perhaps as many as 50% of all studies checked had suspicious results, most notably in the field of psychology. The **replication crisis**[10], the fact many scientific studies have their results taken as true but nobody was able to repeat the underlying experiments, is a huge problem that compounds as certain presumptions get accepted as fact, sometimes out of fear of offending the established scientists.

---

[8] https://news.mit.edu/2015/how-three-mit-students-fooled-scientific-journals-0414

[9] https://pdos.csail.mit.edu/archive/scigen/steeve.pdf

[10] https://simplystatistics.org/2016/08/24/replication-crisis/

Another potential cause is that there's little private interest in funding abstract research, so scientists have to concoct an interesting subject and keep reaching dramatic conclusions, so a government entity keeps giving them money. The perfect example of this is the highly controversial claim that carbon dioxide emissions caused global warming. While that might as well be true, it's impossible to model the Earth's atmosphere to test out how sensitive it is to carbon dioxide, thus violating the core principle of science that all findings should be testable and replicable. There's no telling how much our actions cause global warming and if we're even able to produce a non-trivial drop in global temperature if we were to throw all the money in the world at the problem, but any research that shows so will get as much government funding as needed to figure things out.

The famed **Stanford prison experiment**[11] was done in 1971 by a psychology professor Philip Zimbardo and produced a startling result: we're all equally capable of being cruel. Over the course of six days, the professor split 24 volunteer male students into two groups, prisoners and prison guards, with each group quickly forgetting they were roleplaying and showing unusually vivacious characteristics. The controversy arose some 40 years later when psychologists looked at the original videotapes, in particular, the instructions professor Zimbardo gave to the guards, which essentially amounted to "give me something to work with". Participants also came out and admitted to faking the entire thing by adopting a persona of their favorite film character, but the experiment had already been accepted as gospel and studied in psychology textbooks. The original experiment was funded by the US Office of Naval Research to examine why Navy prisoners become unruly.

---

[11] https://nypost.com/2018/06/14/famed-stanford-prison-experiment-was-a-fraud-scientist-says/

The **marshmallow test**[12], leaving kids with a marshmallow and promising extra rewards if they can refrain from eating it for ten minutes, also made a big splash when it first surfaced in the 1960s, apparently showing key aspects of the child's personality, namely their willpower that enables them to delay gratification. In a curious twist of coincidence, this test was also devised at Stanford University, this time by a psychologist Walter Mischel. Like with the prison experiment, these results were also widely accepted and circulated as true until a repeat experiment done in 2018 failed to replicate the same results. Children who had self-control still did better but nowhere near what the original experiment indicated; even that was made inconsequential by the time children were 15 as their upbringing and environment influenced them much more than previously thought. Circumstances had changed in 50 years, and the repeat experiment included 500 children, ten times more than the original, this time with parents of all backgrounds rather than just those working at Stanford University. Despite being shown as having a minute impact on child's personality, parenting books still parrot the same conclusions from Walter Mischel – children should be taught to delay gratification, and they'll become fully functioning, successful adults. The cause of this could be summed up as "our paper needs to be catchy".

Researchers are often tasked with the requirement of publication known as "publish or perish". That means they simply have to make any kind of finding and corroborate it using whatever data they have handy; in some cases, scientists are urged by superiors to simply keep repeating the experiment until they reach a result that works to support previous findings that brought in funding. The root cause of this might be that humans are simply risk-averse and strive towards security that ultimately leads to stagnation and inability to think outside of established norms. Even when we can repeat the

---

[12]   https://www.theguardian.com/education/2018/jun/01/famed-impulse-control-marshmallow-test-fails-in-new-research

experiment, it's likely our distinct findings can be dismissed as statistical error, so consider the fact that machinery such as LHC is most likely off limits to 99% of scientists in the world and its findings are available only to a small group of carefully selected elite researchers – we've essentially got a priest class that searches for God and presents whatever findings it wants to the general public that can't dispute anything.

Two groups of scientists formed in response to the replication crisis, one trying to mitigate the "publish or perish" syndrome by pre-registering a study for publication – a journal promises to publish a study no matter what its findings end up being – and by sharing findings, software and methods with the public rather than locking them in a vault to discourage cheating. The other group tries to squelch any dissent and rejects any calls to nitpick the data, saying "it's good enough". Dealing with the replication crisis means top scientists have to adopt an attitude of humility and fallibility rather than thinking themselves omniscient; the entire point of science is to prove the older generation wrong while keeping in mind that we're probably wrong too, just less so than they were.

Deep learning is a brand-new paradigm that tries to create the perfect assistant, a way to answer all these and many other questions about the nature of our reality, our origin and our future. In an ideal world, deep learning would provide us with a faster, cheaper and more scalable way to test things out, find answers and have them presented to us in a way compatible with our current scientific knowledge. In reality, there's no telling what could happen as the deep learning technology itself is inherently unpredictable, but there's simply no alternative, at least not one that's in line with the scientific method.

Physicists could have probably just thrown in the towel when first doing the double slit experiment and went fishing, but the insatiable thirst for forbidden knowledge is a fundamental part of the human psyche and what kept them digging for answers. We all crave to know something nobody else does, to find a fundamental rule of some sorts that would allow us to become better than anyone else and surprise evolution, as it were. The question is – what do we do once we find the answer?

Chapter 6 – Evolving the Machine Brain

To approach the idea of unchecked machine evolution brought about through deep learning, let's imagine a pack of cheetahs preying on a herd of grazing antelopes. The two groups have a conflicting survival goal – antelopes want to outrun cheetahs who want to eat them. The evolutionary forces at work will eventually lead to a teetering stalemate in which cheetahs will gradually evolve over hundreds of generations to become smarter, faster and more efficient at killing antelopes, which will also evolve to become smarter, faster and more efficient at *evading* cheetahs; evolution thus becomes an *optimization pressure* brought about through scarcity of resources. Simply put, there's not enough food for everyone, and so a certain number of antelopes will die due to not finding grass, and a certain number of cheetahs will necessarily die due to not being able to catch any prey, but the core members of each population will carry on optimizing.

The primary goal for either group would be survival and any aberration in any individual that makes them unable to survive also eliminates them from the gene pool. If we had to justify such tyrannical evolutionary pressure and present at least one upside, we could say, "Well at least it eliminates the anomalies." In actual examples, this might mean antelopes born without hind legs or with one eye due to a genetic glitch or a cheetah with a crooked spine. Adapting to the necessities of the hunt for cheetahs means developing intelligence or a way to outsmart the antelopes; on the other hand, antelopes develop intelligence in how to evade cheetahs.

Evolution thus weeds out the weakest members but always in an incremental, inter-generational and self-contained manner within the environment. It's not possible to see evolution in action since it moves at a glacial pace, but we can tell it's there because every living thing wants to become better at whatever it's doing to survive with what it has. In other words, evolution in nature is constrained through time, space and limited natural resources. This same pattern repeats across species and along the lines of history as far back as we can see, including how tribes, nations, and empires exist, war against, and absorb one another. Charles Darwin's seminal work "On the Origin of Species" is lauded exactly because it brought to light already seen but never fully understood evolutionary forces; it was as if scientists were sitting in utter darkness and Darwin simply walked in while turning on the light. The theory of evolution seemingly explained everything, but there is a curious exception to the idea of evolution – human civilization.

Rather than competing with one another for resources and opportunistically backstabbing whoever to gain the upper hand, humans figured out that *working together* produces something much more than a mere sum of their individual efforts – a force that can resist evolution and help the entire humanity move in any way they desire. For example, an infant succumbing to a rare genetic disease would have been doomed in any other circumstances, but thousands of researchers dissecting and meticulously studying genes advanced the medicine to the point it can essentially give the baby a second lease of life through genetic therapies or simply by managing the symptoms in perpetuity, as costly as that might be. In a way, we've found a way to cheat the evolution out of its dues and create a much more effortless life for all of us. It's no wonder the elders' rant, "In my day I had it much tougher" because *they actually did.*

The collective sum of struggling against evolutionary forces is how humans built their civilizations, how great thinkers layered one cornerstone after another to allow us all to live easier and more fulfilling lives while the society optimized the incentives given back to worthy individuals. For example, procreation requires a great investment of time, resources and attention in social skills to find a suitable mate to forward the species, so a reclusive inventor that has willingly opted out of the mating dance to focus on his work has no chance of procreation. However, such an inventor is more likely to produce an invention or scientific breakthrough that will advance the entire civilization, which has therefore created scientific awards to give prestige and extra status to this otherwise hopelessly single scientist, helping him attract females.

This is how evolution works and what humans did to bamboozle it, but what would happen if we found a way to create a small machine brain that mimicked the living one, enclosed it in a controlled but unsupervised environment, and let it work out the kinks? We could also scale the experiment by a factor of a million, have the brains automatically tested in regular intervals, shut down those that don't fulfill our expectations, merge those that work, and just keep repeating the process. Such an evolution wouldn't be restricted by space, time or resource scarcity, allowing it to blossom beyond our wildest expectations.

The power consumption would be minimal, the storage needed would be laughably small, and the overall investment would also be close to nothing, *especially* if the creators found a way to profit off the intermediary evolution stages of such digital life. For example, one such digital being could be optimized to recognize images and automatically tag people uploading selfies to their social media account; the initial investment in its training and maintenance would be recouped by leasing its capabilities to anyone who can afford it at 100 times the markup.

At first these digital brains might not be any smarter than a cockroach, which knows to distinguish between dark and light and skitter away when exposed to the potential predator, but they would eventually evolve into something as smart as a sparrow or a raven, which can see shapes, colors and understand much more of their surroundings than a cockroach. This is the scary prospect of unregulated, unsupervised evolution done in cyberspace, a sliver of digital space contained on a hard disk in a warehouse somewhere near the Arctic Circle. The cyberspace would represent a simplified version of our universe, unsuitable for organic but perfect for digital life.

This digital life inhabiting the cyberspace would develop **artificial intelligence** (AI), a way of grasping the physical world in real time, that would at first be **narrow AI**, capable only of the simplest tasks, such as comparing two items or two colors. Eventually, the narrow AI would become **general AI**, as smart as a human, though this would require a **quantum leap** in their efficiency, an unknown breakthrough nobody can predict or control. For now, we've still got time, and a machine capable of human-like thought is still far off in the future, but the scary thing is what would happen afterward.

Since we examined how humans managed to combine their talents into a civilization, and created a force that can rival that of evolution itself, a network of general AI brains would quite possibly be capable of creating a **super AI**, an unstoppable digital entity with godlike powers that would also have the ability to take over the internet and every device connected to it, doing with humans as it pleases – such super AI might be able to solve previously unsolvable conundrums, such as the double slit experiment, but could also make us its pets and expose us to hardships of evolution all over again. For now, the super AI is far off in the future, but the building of narrow AI neural networks continues at a steady pace with each new academic paper, such as the ones presented below.

Distinguishing between music genres

It's in the nature of artists to defy expectations and invent something brand new, whether it's shocking, inspiring or awe-inducing, which is why trying to classify music genres can be a headache. With painters, we can at least use their painting methods to pin them down to a general genre, but with musicians, anything goes – drums, whistles, air horns and slamming two trashcan lids together can find their place in a single segment and actually sound good. Naturally, as soon as that kind of sound becomes popular enough, there's bound to be copycats adding their own flair and further fracturing the genre. In a sense, only those who've heard the originator of the genre can correctly hear the underlying beat to identify the genre and ignore all the later additions by copycats.

Perhaps the difficulty with music is that hearing is much more fragile and prone to losing access to certain frequencies, causing each person to literally hear a slightly different thing, which is especially notable with kids being able to hear all sorts of buzzing and chirping sounds adults can't. So, when scientists ran out of ideas, they threw a neural network at this music genre classification problem, and it actually stuck.

In 2016, two scientists thought of creating a neural network that can correctly identify a musical genre, not merely as an overall estimate of the entire song but as a graph that identifies exact moments where a song sounds more jazz and when it flips to pure rock. In this way, the neural network could present a song as a combination of distinct musical identifiers but also help us create a branching diagram of musicians that can be traced back to the originator to help us figure out that new musician and where he got his inspiration. Named DeepSound[13], the neural network was featured at Warsaw 2016 Braincode hackathon and actually won first place.

The two scientists behind DeepSound first tackled the theoretical part of the problem by contacting friends at musical institutes and asking them to help with identifying unique, basic frequencies that mark the genre as a whole. The solution was in using a **spectrogram**, a graph that shows the intensity of a signal with time, and is used in all sorts of signal analysis. It turns out that

---

[13] http://deepsound.io/music_genre_recognition.html

spectrograms of classical music compared to those of jazz have distinctly different peaks and lows, clearly showing basic frequencies and additions. The peaks and lows can be even further simplified as vector data that shows genre probability distribution. Now that we've got music broken down to its components, the neural network can be fed data to recognize and label.

The authors reasoned that a rock song would mostly contain rock features, so they figured out that arithmetic mean of these genre probability vectors would strongly indicate what the song genre is. This means the vectors are summed and divided by their amount; the result of mathematical operations on vectors is also a vector. Training was done on 700 music samples and testing on a further 300, with the neural network correctly categorizing 67% of *music content*, not merely songs as a whole. This number doesn't seem impressive compared to prior music genre label generator, but the authors did give themselves an as-of-yet indomitable task and fared pretty well on it compared to a random guessing model that had only 10% accuracy.

Spotify did something similar with deep learning but originally relied mostly on aggregate user behavior to determine if songs are similar – if two users show interest in what Spotify definitely knows are rock songs, then other music they listen to on Spotify is most likely rock. Scale this up to 1,000 users and numbers start to converge, giving us a solid prediction model that can be used to make recommendations to brand-new users. This approach was termed "collaborative filtering" and doesn't really have much to do with neural networks, not yet at least.

Collaborative filtering is a universal approach to categorizing all sorts of data, and the exact same model can be used to recommend books, shoes, video games and whatever else. There is a problem though – collaborative filtering strongly favors popular items and discourages discovery. Niche musician with great jazz skill on Spotify? No way for anyone to find him except by pure chance or if he invests thousands into advertising, which does benefit Spotify but requires the musician to actually sell his music to get the money, which would require exposure – it's a Catch 22 or to use a fancier term "cold start problem".

Popular items get much more traffic and consequently usage data to classify them than niche items, which is why collaborative filtering often results in boring, predictable and bland recommendations – not the kinds of feeling any company would want to be associated with. With only the biggest bands and musicians recognized, everyone else gets squeezed out, and Spotify turns from this hip, cool place to find new music to just another mainstream music outlet. So, when Spotify ran out of ideas, they started tinkering with a neural network.

Spotify first thought of expanding their data gathering efforts, so they acquired 13 companies from 2013-18, most dealing with tagging, sharing and exploring musical genres. There's a lot of data associated with every musical piece, such as when it was made and what instruments were used. Some of these can be guesstimated, but for things such as mood of the song and lyrics, there's no clear solution, so Spotify hired three researchers to figure out how to connect the dots. The result of their research was a paper "Deep content-based music recommendation"[14].

The idea in the paper is that of what the authors termed "latent space", a way of presenting known data on songs as 2D vectors and correlating them; songs close together probably sound similar, but the novel addition to this concept is that *users* were represented as vectors as well based on known information on them, such as age, gender, ethnicity, etc. The result is that users could be plotted in the latent space alongside songs and if the two match, we might have just found a way to recommend new songs to users. Using neural network means songs can be placed in the latent space with a high degree of certainty as we train it to extract musical features.

---

[14] https://papers.nips.cc/paper/5004-deep-content-based-music-recommendation.pdf

Songs were first cut up in three-second pieces, with the neural network predicting each segment's qualities and averaging out the values. Once Spotify saw these guys were on to something, they were given full access to the Spotify musical library, in particular, 1 million songs and 30 seconds from the middle of each. Researchers aimed at a neural network that would closely resemble ones used for image classification that get better at recognizing features as they're trained. In this case, each node in the network eventually learned to pick up a single feature of the song, such as ambient noise, bass drums, Chinese words or vibrato singing. Funnily enough, Armin van Buuren was found to have a filter all to himself.

The results of this research are difficult to quantify since there's no objective standard to measure the similarity of songs to one another, but it's yet another step towards finding a comprehensive neural network solution for how to find the best kind of music for everyone. Spotify was bought by Microsoft in 2018 for $41.8bn, showing just how much money there is to be made in classifying music genres.

"Review-Driven Multi-Label Music Style Classification by Exploiting Style Correlations"[15] is another attempt to classify music, this time by eschewing the entire hear-music-analyze-sounds approach. The authors of this particular paper realized that there are plenty of humans who know their particular music better than anyone else; it's only a matter of gathering these publicly available gems and collating them into multi-labels for each song. The authors also understood that each song could have more than one genre, which is why the multi-label idea would fit nicely.

The paper gives examples of reviews and in the case of *Mozart: The Great Piano Concertos, Vol.*1 two such reviews are, "I've been listening to *classical* music all the time" and "Mozart is always good. There is a reason he is ranked in the top 3 of lists of greatest *classical* composers." As users mention instruments, pitches,

---

[15] https://arxiv.org/pdf/1808.07604.pdf

ambiance and other sounds, the neural network starts getting more and more descriptive labels, and the third review on the same piece is, "The sound of *piano* brings me peace and relaxation." Every music genre can thus be represented by a simple equation: Mozart = piano + classical music.

Tor deanonymization attack

Tor (The Onion Router) is a way to theoretically browse the web while remaining anonymous, which could be useful for whistleblowers and reporters in no-speech countries. Tor's main gimmick is that a volunteer network of relay nodes passes network traffic coming from the entry node between themselves until it's seemingly anonymous and then forwards it to the exit node. Anyone trying to distinguish entry from exit nodes would feel like peeling an onion in search of the center – it's all just intermediate layers hence the onion reference. Note that Tor is the name for the network itself that can be used by the Tor browser, Tor instant messaging clients, etc.

Main weaknesses of Tor are that it relies on a healthy and diverse network of relay nodes, with a well-funded actor, such as a government intelligence agency possibly being able to supplant enough volunteer nodes in a region with their own machines and just follow the traffic along as it bounces within the network. Users of Tor could also blow their own cover if they're customizing the browser to values that fingerprint them; browsers report their window size to the website to get appropriate image sizes so a Tor user that's resized their browser window from the default 500x1000 pixels to full monitor size could be tracked just as easily[16] as if they had spyware on their machine. Once deep learning gets involved, debunking Tor users becomes trivially easy.

---

[16]   https://tor.stackexchange.com/questions/16111/is-manually-resizing-the-tor-window-dangerous

"DeepCorr: Strong Flow Correlation Attacks on Tor Using Deep Learning"[17] shows a technique known as **flow correlation**, comparing anonymous network traffic flows to match users to traffic patterns within the Tor network. Flow correlation tracking techniques already exist, but DeepCorr combines them with deep learning to drastically increase tracking efficiency and reliability. By scraping a mere 900KB of Tor data, DeepCorr can match anonymous data flows with 96% accuracy compared to conventional deanonymizing RAPTOR technique that would only have 4% accuracy with the same dataset and would take up to 100MB of data over five minutes of uninterrupted tracking of 50 Tor nodes to achieve the same accuracy.

Anonymity always comes with a tradeoff, but there's one thing internet users would never give up – speed. There could be an indeterminate number of bounces within the network for any given connection attempt, so Tor doesn't obfuscate certain characteristics of network packets or try to hide *when* they were sent, all of this in order to minimize latency. This opens Tor up to DeepCorr that matches packet sizes and timings to individuals within the network, something only possible at scale with a neural network.

Tor has long been thought to have inherent resistance to flow correlation attacks due to sheer size (2 million nodes online at any given time) and a large discrepancy between user bandwidth requirements and Tor's volunteer relay network sharing their own bandwidth. This tends to lead to congestion akin to traffic jams until more relay nodes come online to suddenly increase network capacity, causing what's known as **network jitter**, a phenomenon where packets get congested and slightly delayed on their trajectory. This causes a lot of noise that confuses traditional tracking mechanisms, but DeepCorr is designed to match Tor's hectic ecosystem and learn to recognize traffic patterns even if entry and exit nodes are unknown or unknowable.

---

[17] https://arxiv.org/pdf/1808.07285.pdf

Authors of the paper tested DeepCorr by finding the top 50,000 websites as ranked by Alexa, browsing them using Tor and training the neural network with half that dataset using a single GeForce GTX TITAN X 12 GB (priced $600-1,000 depending on the version) for a day; the other half was used to test DeepCorr. The authors estimate that retraining DeepCorr once a month would suffice to keep it on par with any updates to the Tor network protocol. The dataset was made by opening up to ten concurrent Tor connections using Tor and Firefox browsers within standalone virtual machines, capturing the outgoing traffic and forcing the traffic through a proxy server set up by the authors where it was also captured.

The conclusions were that DeepCorr's accuracy didn't significantly diminish with the passage of time up to a month but did experience degradation in confidence after that. Flow lengths were also positively correlated with DeepCorr's efficiency, and the more data captured, the better the results – though authors do note this requires exponentially more storage and network capacity, leading us back to the question of a well-funded adversary attacking the Tor network. Finally, DeepCorr can just as easily track users performing proxy cyberattacks, the ones depicted in the movies as a series of lines connecting the dots on the globe as panicked operators count the time down.

Playing poker

Poker is a gentleman's game that involves skillful estimation of probability and nuanced understanding of nonverbal cues and ticks to play towards one's outs. Just like with any other activity, humans used to dawdle in neural networks, turned everything upside down, and conclusively solved poker. Time to put on a poker face, wear sunglasses indoors and go all in on "Approximating Poker Probabilities with Deep Learning"[18], a paper that presents an alternative to **Monte Carlo** tree search.

---

[18] https://arxiv.org/pdf/1808.07220.pdf

Monte Carlo simulates all the outcomes of a game with branching choices such as poker by going through them one by one, noting the final outcome – win or lose – and updating the count at the start of the branch. This comes out to millions of outcomes that would take a supercomputer to crunch in real time, but a neural network can do just as much in a computationally lightweight manner while also including opponent modeling to account for behavior biases all humans have. Some people like risk and others play safer.

Poker probabilities gotten through Monte Carlo are still just an approximation since there are unknown variables that affect the game, but running the simulation 1,000 times should present converging probabilities that would point to our hand's true strength with a 2% error margin. Even then we run into trouble as one Monte Carlo simulation took the author 0.46563 seconds, which means running it a million times would take close to 129 hours and even that's not enough to cover all the branches.

The neural network presented in this paper suggests approximating Monte Carlo, which would mean approximating an approximation; this could raise some eyebrows, but as long as we take a large enough Monte Carlo sample size, the dataset will remain stable. As a result, the neural network was able to run 600 times faster than Monte Carlo and take up only 8.4KB of memory while guessing which hand would be the winning one in 79% of cases and which would be a tying one in 95% of cases with a 5% margin of error in each.

Person recognition

Finding Waldo was once a pleasant pastime for humans, but neural networks have made picking out any given persons out of the background an easily solvable challenge. "Person Search by Multi-Scale Matching"[19] addresses the issue of multi-scale matching, meaning that the person we're looking for can be at any given angle

---

[19] https://arxiv.org/pdf/1807.08582.pdf

and distance from the camera and the neural network still has to figure out if that's whom we're looking for across all offered images. The biggest advantage of a neural network compared to humans doing the same task is that it has no bias towards bigger images with more details while avoiding low resolution, blurry or awkwardly shot images as we do. The neural network managed a mean average precision of 87.2% even in images where the person was within an area as small as 37×13 pixels.

Tool breakage

Industrial tool breakage is a serious issue since tool bits endure tremendous load for quite a while before suddenly shattering without any warning; tool owners want to squeeze out maximum performance to increase profit margins, but breakage causes delays and additional maintenance costs. The research paper "Tool Breakage Detection using Deep Learning"[20] tries to solve this problem by employing neural networks that use deep learning to find out the exact moment when the tool is about to break, eking out maximum durability out of any given tool piece.

One curious feature of this particular challenge is that training a neural network to predict the breaking point of any given tool using large datasets isn't feasible as the margin for error is unrealistically large. The idea in this paper is to track power consumption of a machine as it's running and let the neural network find a rule as to tool breakage point and power consumption of the machine as a whole. Workers experienced with a certain machine are in a sense sensitized to the sound of its operation and can hear the slightly different sound of a worn tool bit, but the environmental noise often impedes their judgment; an ideally modeled neural network would be able to predict breakage reliably regardless of noise levels. The authors conclude that their approach can achieve 93% prediction

---

[20] https://arxiv.org/pdf/1808.05347.pdf

precision with a comment that the model has plenty of room for improvement.

Taxi fleet advance positioning

We've all encountered Schrodinger's Taxi – a taxi vehicle exists at our location in a state of quantum superposition that is collapsed only when we look around in dire need of transport. That's not the actual reason why we often can't find a taxi, but it might as well be since people travel for all sorts of reasons that are hard to contextualize and predict, causing taxis to flock to seemingly random locations for a chance of getting a fare. Make no mistake, cracking the code to travel habits of people on a large scale would make gazillions to the lucky taxi company, which is why "Combining time-series and textual data for taxi demand prediction in event areas: a deep learning approach"[21] uses a neural network to do just that.

The two most common travel patterns for humans are habitual, meaning the same person regularly goes the same route at similar times, and causal, meaning the person is traveling between two logically connected places such as from an airport to a hotel. In a place such as New York, there are so many interesting events that being able to predict accurately and model travel patterns that don't fall into any of the above two would lead to significant gains in efficiency. The main problem is what the taxi industry calls "demand surges", the notion of being able to anticipate higher demand at a particular location and preemptively send vehicles there to grab fares before anyone else. The problem is that traditional polling and data gathering methods to figure out demand surges are too slow and don't cover unstructured text patterns commonly used by young adults to convey information, so those are exactly the problems this paper will try to solve.

---

[21] https://arxiv.org/pdf/1808.05535.pdf

By analyzing data on 1.1 billion New York taxi trips made in 2009-2016, and combining the results with semantic analysis tools that try to figure out author intentions from publicly written texts such as tweets, a neural network can first derive habitual and causal trends before removing them from the overall data and focusing on the rest; those are demand surges that we plug back into the semantic analysis and end up knowing more about travel habits of people than they know themselves. It's as easy as pie.

Semantic analysis tools also process the text data by stripping all HTML tags, removing inflectional endings to get the word root, making text all lowercase, and ignoring very frequent and very rare words such as articles or misspelled words. Each word can then be turned into a one-dimensional vector, making each sentence a straight line that indicates direction and travel distance. Vectors can be placed within a 300-dimensional space that indicates word similarity and closeness in meaning, such as that a vector "female" and a vector "king" converge with the vector "queen".

Two venues were picked for analysis – Barclays Center sports arena, the home of the Brooklyn Nets, and Terminal 5, a three-story concert venue. All taxi pickups within 500 meters from either of those were taken into account for a total of 1,066 events, and the weather data was independently assessed for each day, such as precipitation, temperature, etc. The neural network's job was to analyze roughly half of all available data with the other half being given by the scientists as a challenge afterward.

Results showed that a neural network using all this contextual data was able to reduce mean absolute percentage error (MAPE) by almost 30% compared to Gaussian process, another state-of-the-art prediction algorithm. The paper concludes that data mining publicly available information sources, and contextualizing them using a neural network, can be used by taxi companies or any other transport competitor, such as Uber, to gain the upper hand by positioning vehicle fleets the night before demand surges. The authors of this paper made their work open source so anyone can profit from it.

Power grid management

The light bulb flickers and the PC beeps, restarting in a blink of an eye – great, there go two hours of unsaved work. For those living outside of metropolitan areas, this kind of experience caused by power fluctuations is a common occurrence they learn to tolerate, but there's very little the power company could have done to stop it. The exact cause of the power fluctuation is called "peak load", a moment when for whatever reason all customers in a certain area start drawing more power than they usually do, causing certain gadgets to stop working.

The way a power company currently supplies consumers with life-giving electrons is by calculating what each household spends at its minimum, producing slightly above that while keeping a certain amount of energy in storage and selling the rest to other power grids in need. The sale and purchase of power are done on a power market on an hourly or daily basis, with price oscillating depending on regional demand. Being able to predict incoming peak loads correctly and buy up power right before it happens can potentially save a power company millions and in a sense also save its users' unsaved work. Peak loads are always brief moments, and a power grid can, in theory, withstand a prolonged peak load but in practice generally experiences a system-wide failure.

Power grids across an entire continent can be interconnected and act as a joint power transmission vessel that keeps power levels stable no matter the peak loads in component parts. This actually caused a problem in March 2018 as Serbia and its rogue province of Kosovo entered a power grid dispute – the latter drew 113GWh more power from the former than it was able to generate, causing a cascading effect that impacted the entire EU power grid. In the end, electric clocks across the continent fell behind six entire minutes[22] since they

---

[22] https://www.theguardian.com/world/2018/mar/08/european-clocks-lose-six-minutes-dispute-power-electricity-grid

use the power grid's 50Hz frequency to measure time; the power drain caused the frequency to dip slightly below that. Though this is an outlier case, regional power suppliers benefit from understanding power fluctuations and peak load timings so they can prepare extra support staff, acquire power in advance at a cheaper price, etc.

Electricity purchase timing

The research paper "Deep Learning for Energy Markets"[23] goes on to evaluate power price data on a daily and hourly basis when it comes to 4719 power-generating nodes in the US and what they had to do to keep the output stable in 2017. The dataset consisted of prices, weather, demand, and output and was given to the neural network to correlate them correctly. The results showed that high power demand doesn't necessarily involve high prices, with the neural network precisely laying out the cyclical nature of peak load timings though it generally underestimated peak prices by some 20%. The authors of the paper concluded that this approach has "better accuracy than traditional time series models."

Detecting electricity theft

In the US, power theft drains about $6bn a year; in some countries, all revenue a power company gets is drained by the thieves. The problem is that any losses incurred by the power company eventually get shifted onto honest customers in the form of various fees and steady price increases. Smart Meters are the newest fad when it comes to precisely detecting power consumption, but they still don't solve the problem of advanced electricity theft, i.e., cyberattacks on the Smart Meter or the power grid. Since Smart Meters are an electronic device, they're prone to being hacked or reprogrammed without the power company ever figuring it out, but with the help of neural networks, it might finally be lights out for the electricity thieves.

---

[23] https://arxiv.org/pdf/1808.05527.pdf

"Deep Recurrent Electricity Theft Detection in AMI Networks with Random Tuning of Hyper-parameters"[24] uses 200 customers' consumption data over the course of 107,200 days to detect fine-grained usage patterns and discover what tampering with the meter or the power supply network looks like. Trying to accrue historical consumption data for any given client doesn't work for those just joining the grid, but a neural network can tweak its own procedure to improve over time, distinguishing between malicious and honest customers with a 93% detection rate and a 5% false-positive rate.

Driver drowsiness detection

"EEG-Based Driver Drowsiness Estimation Using Convolutional Neural Networks"[25] suggests monitoring the driver to immediately recognize not just when he's about to fall asleep, but distraction and speeding as well to take immediate action such as sounding the alarm or slowing down. The most common approach in this kind of detection is to place a windshield camera that tracks and analyzes his head and eye movement in real time to instantly react to any untoward behavior; a camera can also be placed facing the road to spot when the vehicle starts veering outside the lines and react. Both approaches suffer in bad lighting and weather conditions, annoying the driver and ironically serving as a distraction that will cause an accident rather than stop it.

The alternative is to have the driver wear some sort of contact equipment, such as sensors, that accurately measures physiological signs, such as brain waves, but random fluctuations in bodily response can also interfere with the readings. Sensors themselves are not that comfortable to wear, again causing distraction for the driver, but that's what the authors of this paper went with, opting for brain wave sensors.

---

[24] https://arxiv.org/pdf/1809.01774.pdf

[25] https://arxiv.org/pdf/1809.00929.pdf

16 healthy persons were chosen to drive a vehicle using a virtual reality setup for 60-90 minutes during the afternoon, when the urge to take a nap is the strongest. The vehicle traveled 100 km/h (about 60 mph) along a monotonous vista, but there was a sudden random change in traffic and drivers were instructed to steer away immediately. Their reaction times were recorded to reach a drowsiness index, and compared with their brain wave activity, reaching a 63.79% correlation coefficient that can roughly equate with accuracy.

Neural network architecture search

"Neural Architecture Search: A Survey"[26] focuses on building such a neural network that can evolve on its own, create offspring neural networks and find the most optimal neural network layout for any given problem. The neural architecture search (NAS) is in essence automated machine learning that involves three parameters: search space, search strategy, and performance estimation. The only human involvement is setting these three parameters and assessing the results, though the paper laments the fact humans introduce their bias by limiting the neural network's search space.

Search space refers to the number of architecture models we'll let the neural network assess and recombine to reach something genuinely new and better than any of them. A novel proposition when it comes to search space is the introduction of "cells", hand-crafted neural network components that can preserve the data multidimensionality or be made to reduce data to a vector. Cells can be transplanted to another neural network or stacked on top of one another for unparalleled performance gains, but the notion is to let the neural network combine them in arbitrary ways until there is significant progress.

Any number of search strategies can be employed by the neural network, though the performance of some of them declines as the

[26] https://arxiv.org/pdf/1808.05377.pdf

network scales, making them unfeasible. For example, reinforcement learning search strategy[27] used by Zoph and Le in 2017 required 800 computer graphics cards running for three-four weeks. However, evolutionary methods were used all the way back in the early 90s and allowed considerable search space exploration with low-key resource consumption. The main idea behind the evolutionary method is that the parent neural network runs for a while until it reaches a certain architecture, notes its efficiency on the task and then passes the layout on to its offspring, a brand-new neural network that now has some idea on how well certain cells perform on the task and can weight them accordingly. By gradually weeding out certain cells and their layouts the offspring can eventually reach the ultimate neural network architecture for the given task. Now, all that's left is performance estimation.

Despite cutting corners with cells and using the evolutionary method to increase scalability NAS still incurs significant resource drain when it's time to test the proposed neural network architecture, prompting the authors to resort to *performance estimation* that is, in essence, lo-fi measurement. The trick is that the estimate can't oversimplify or the results will be no good, so the authors resort to merely ranking resulting neural networks based on their performance, with the idea that we don't need to know how good the best one is, only that it *is* the best one of the bunch.

The authors conclude that NAS fared well, but ranking performance is nonetheless difficult due to the lack of common benchmark standards in deep learning. Another remark is that NAS doesn't really reveal why a certain architecture performs in a certain way and that understanding cell groupings (also called "motifs" in the paper) would provide insight into how neural networks work.

---

[27] https://arxiv.org/pdf/1709.07417.pdf

Noise recognition

The 2018 research paper "Noise Adaptive Speech Enhancement Using Domain Adversarial Training"[28] looks at speech recognition neural networks trained using deep learning and their ability to handle audio sources with additional background noises not encountered during the training phase. Rather than compiling the list of every noise ever created, the paper suggests using domain adversarial training (DAT) to train two additional subroutines of the neural network: a discriminator that tries to determine if the noise is coming from the original audio source and a feature extractor that tries to produce the best noise to confuse the discriminator. By pitting a discriminator against a feature extractor over an arbitrarily long period of time, the DAT technique allows the evolution of a specialized discriminator module that can later be merged with the audio recognition neural network for 26-55% better performance in devices such as cochlear implants and speech-to-text software. This kind of adversarial training is typical for neural networks as it allows scientists to leverage the blazing speed of computers for efficient learning compared to what would happen if they were manually fed examples.

Scene recognition

"From Volcano to Toyshop: Adaptive Discriminative Region Discovery for Scene Recognition"[29] looks at an image classifying neural network that is first trained to label objects in the scene, then label the scenery itself, and finally contextualize both to arrive at a description of the location, for example, "art school" or "campsite". The idea is to train the neural network to recognize certain objects as highly deterministic of the location, for example, finding a tent in the

---

[28] https://arxiv.org/pdf/1807.07501.pdf

[29] https://arxiv.org/pdf/1807.08624.pdf

image strongly suggests it's outdoors. Such neural networks already exist but are computationally demanding both in training and in operation; deep learning allows the operator to set an arbitrary number of signifiers in the image to achieve scalability.

Decrypting hidden messages

Steganography is intentionally hiding information in other data, and steganalysis is a way to unveil such hidden messages, both of which neural networks do much better than humans as revealed in "Invisible Steganography via Generative Adversarial Network"[30]. The idea is to train two separate neural networks, one in hiding the information and the other in unveiling it, by pitting them against one another. The test given was to have one neural network analyze the cover image, find the most suitable pixels, successfully hide a gray image into a color one, and then send it to the other neural network for analysis. When working in tandem the two networks allow, for example, the headquarters to transmit a secret message encoded in a plain image through public channels to operatives in the field who can decode it using the other part of the duo.

Automatic question generation

High school quizzes taught us many facts, such as that mitochondria is the powerhouse of the cell. Preparing the quizzes required a lot of delicate but hasty work that could go wrong in any number of places, making students feel cheated out of a good grade due to a malformed question. With the help of deep learning, we might be on the brink of an age where questions and answers are unmistakably created from swaths of text by a neural network, the powerhouse of the learning process.

"Improving Neural Question Generation using Answer Separation"[31] looks at automatically creating questions and answers out of any

---

[30] https://arxiv.org/pdf/1807.08571.pdf

[31] https://arxiv.org/pdf/1809.02393.pdf

amount of text, ranging from single sentences to large paragraphs. The goal isn't to have the neural network create a rough draft of the quiz but an actual workable version from scratch that doesn't need any proofreading or editing. This is done by having the neural network identify and mask the answer with a token signifier and semantically conclude the correct word sequence and pronouns.

For example, the sentence "John Francis O'Hara was elected president of Notre Dame in 1934" contains three questions: who (John), what (president) and when (1934). By masking either of the three answers, we teach the neural network how to pose a question and test it against the masked answer; by adding an attention mechanism we give higher weight to keywords and information tidbits, mimicking the way humans parse questions and retain knowledge. This approach allows extraction of maximum value out of the same text but also acts as an anti-cheat measure during the quiz itself since students can no longer copy an answer from someone else.

The neural network was tested using 23,215 text samples originating from 536 text sources with about 100,000 questions and answers created manually for actual quizzes. The results showed that only 0.6% of questions created by this neural network erroneously revealed the entire answer and 9.5% gave a hint as to what the answer was, both being common weak points of neural networks dealing with creating quiz questions. "What", "how" and "who" were the three most common pronouns that the neural network guessed correctly, though the average precision for other question types wasn't so stellar; the authors attributed this to 55.4% of all questions having "what" and other pronouns not being nearly as represented in the training dataset.

3D visual reconstruction of 2D objects

Great painters know how to use shades and perspective to make the canvas a genuine window into another world[32]. Even though on some level we know that this kind of painting is an illusion, our brain snaps the 3D picture together and presents it as real, filling in the blanks using what it knows about the outside world. It turns out neural networks are capable of something similar and can reconstruct the unseen side of an object based off of one of its 2D views.

"Deep Learned Full-3D Object Completion from Single View"[33] is a joint USA-Italy research paper looking at how pixels can be turned into **voxels**, best described as volumetric pixels. The main goal of this study is to minimize the number of perspectives needed to complete a 3D object, possibly to be used with robots on a tight computational budget moving through an environment and interacting with real objects. Since the authors needed at least one view, they decided to go with that and ended up actually succeeding.

The neural network was trained using 5,000 model presets from CAD, a popular modeling program, each model having eight snapshots taken from different angles to arrive at 40,000 challenges. All models were at a 30x30x30 resolution, which presented a challenge when it came to preserving all the nuanced features of a model, but the neural network managed to achieve impressive results nonetheless, restoring 92% of the original model.

Rain streak removal from an image or video feed

Setting up a remotely accessible outdoors camera can sound like a fun experiment; that is until the rain falls and rain streaks make it

---

[32] https://uploads1.wikiart.org/images/albert-bierstadt/the-falls-of-st-anthony.jpg

[33] https://arxiv.org/pdf/1808.06843.pdf

seem like we're peeking through a white curtain. The actual cause of this effect is that raindrops have a high velocity while reflecting light, causing white streaks to appear on any image capturing device. This devalues all other machine learning processes that depend on image processing, such as facial recognition, and thus removing rain streaks becomes a top priority project. There currently exist several de-raining programs, some of which use machine learning, but they generally ruin the picture quality either by blurring the background or by ruining image contrast. "Rain Streak Removal for Single Image via Kernel Guided CNN"[34] suggests using a neural network called KGCNN to clean the images up with the goals being to preserve as much picture quality as possible and remove rain streaks from the image in a computationally lightweight way.

KGCNN would exploit a known property of raindrops, that being that they induce a small but perceptible amount of motion blur. Knowing the general direction of a raindrop, namely that they tend to fall down, can help us build KGCNN that will deconstruct any image feed into a background and texture layer, with the goal being to identify motion blur in the latter and use this knowledge back on the composite image to block out the rain streaks. The background layer contains everything except the rain, and the texture layer contains only the raindrops, making it easy to identify at a glance if KGCNN works as expected.

Computer systems intrusion detection

In sci-fi books such as William Gibson's *Neuromancer*, hackers plug into the internet via a literal plug inserted into the base of the skull and jockey the software around; intrusion detection is done by an AI called Black Ice that guards proprietary cyberspace and tries to fry the hackers' brains. Hackers and intrusion detection exist right now in a much more prosaic manner, but neural networks promise that at least that latter part is about to become much more interesting.

---

[34] https://arxiv.org/pdf/1808.08545.pdf

"Statistical Analysis Driven Optimized Deep Learning System for Intrusion Detection"[35] investigates the rising threat of intelligent malware and hacking attacks that might jeopardize banking systems, power grids or hospital record databases. It's not just that everything is networked, but the sheer size of such networks makes updating a nightmare and increases their attack surface, leaving them exposed to any hacker with an opportunity; it's as easy as walking to one of these terminals connected to the **intranet**, the internal network, and popping in an infected USB. It's probably how WannaCry ransomware hit 16 UK hospitals in May 2017, locking all medical files behind a paywall and threatening deletion unless $300 in ransom was paid in Bitcoin[36].

This research paper shows a scalable, lightweight way to keep huge networks secure, regardless of whether their components are updated or not, by using neural networks that sift through a massive volume of data to anticipate intruder behavior and deny them access. We know from other scientific areas that neural networks can achieve near-human performance in cases of person recognition and 3D object reconstruction, so it's of great interest to find a sustainable, cheap alternative to antivirus programs and outdated control access routines that cause endless grief to support staff, such as usernames and passwords.

The neural network assigned to network security does data preprocessing to remove outliers, feature extraction to find commonalities between users, and classification to distinguish between benign and malign users. In general, intrusions come as: probing that scouts the target network for weaknesses and open ports; denial-of-service, which serves to incapacitate target network and estimate its capabilities; user-to-root that is meant to gain root

---

[35] https://arxiv.org/pdf/1808.05633.pdf

[36]   https://www.theverge.com/2017/5/12/15630354/nhs-hospitals-ransomware-hack-wannacry-bitcoin

access; and root-to-local that is meant to perform operations on a local machine after root access has been gained.

Neural network was first trained using 125,973 samples belonging to any of the four intrusion categories and tested with another 22,544 samples, resulting in an accuracy of 77.13% for probing attacks, 97.08% for denial-of-service, 87.10% for user-to-root but only 11.74% for root-to-local. These numbers imply that all security measures should focus on preventing access to root systems, which are those that can issue commands to subordinate machines, including the entire network or individual terminals; once the root has been hijacked, every compromised machine, or even the entire network, is decisively in the hands of the attacker, as seen with WannaCry ransomware.

Logical thinking

Logic sets man apart from amoebas and lets us see both ourselves and amoebas with a high degree of certainty. The ability of logical reasoning derives from a symbolic representation of the world and is the most yearned-for faculty scientists want their neural networks to have. It's not just any kind of logic, though, but a special kind called ontological logic that deals with how we came to be and what ties us together. Ontological logic can be applied to countries, humans, animals, trees, rocks or any other conceivable material or metaphysical entity to grab the fabric of time and unravel it back to its starting point. As neural networks get set to compete against humans in all fields of life and science, they'll slowly but surely learn how to navel-gaze just like the rest of us with the free time to do so.

"Ontology Reasoning with Deep Neural Networks"[37] looks at how a neural network that's been given facts about people draws conclusions about their relationships; for example, two separate individuals that happen to be parents to the same person must be

---

[37] https://arxiv.org/pdf/1808.07980.pdf

related as well. The same logic is then applied to cities, provinces, countries themselves, etc., to let the neural network learn new things about the world and update its own information database. This is in part how Facebook's "People You May Know" feature works – figuring out the origin of a relationship often reveals intimate details that might have gone unnoticed even by the people involved. We do have such massive information databases, but something like Wikipedia uses throngs of unpaid volunteers bitterly bickering for weeks about interpunction in obscure articles to provide the bulk of text used by the public; this kind of a neural network could be used to either test a wiki or build a brand-new one.

Testing was done with two information databases, Claros and DBpedia. Though not the same size, the neural network learned how to correctly interpret objects, data and relations between them with a 99.8% accuracy. The authors then decided to up the challenge by removing a random fact from each database and replacing them with a diametrically opposite version of themselves, meaning "man" might have been replaced by "woman" etc. This created a conflict we would call a **paradox**, a statement that appears both true and false at the same time, but the neural network managed to resolve on average 92% of all conflicts. The authors did note that facts presented with less than 100% certainty in the information database did throw a wrench in the neural network's spokes.

Fooling the smart machine

Videos on the Open AI blog[38] examine how image classifying neural networks can be fooled by presenting a printed image of a kitten digitally altered to contain blocky aftereffects. The image is clearly recognizable by the human eye, but a neural network sees a desktop computer under almost all angles and zoom factors, persisting even when the image is rotated or moved aside. The related 2018 research

---

[38] https://blog.openai.com/robust-adversarial-inputs/

paper titled "Synthesizing Robust Adversarial Examples"[39] investigates the idea of turning 2D images and 3D-printed objects into a source of headaches for the neural network through the use of EOT (Expectation Over Transformation) algorithm that persists even when the image or object are rotated, filmed under different lighting or shown zoomed in or out. In the example shown in the paper, 8/10 images of a 3D-printed turtle were identified by the neural network as a rifle and the rest as "other".

The 2015 research paper titled "DeepFool: a simple and accurate method to fool deep neural networks"[40] describes an algorithm that adds minimal perturbations to any given picture to have the image recognition neural network see it as something else entirely, an example shown being a whale recognized as a turtle. The DeepFool method is then compared to similar perturbation algorithms in terms of cost, speed, and intrusion on the original image, discussing its use in understanding the architecture of any given smart machine and how to optimize the attack.

---

[39] https://arxiv.org/pdf/1707.07397.pdf

[40] https://arxiv.org/pdf/1511.04599.pdf

Chapter 7 – The Future of Deep Learning

"Deep learning, deep change? Mapping the development of the Artificial Intelligence General Purpose Technology"[41] examines business applications and regional clustering of deep learning by comparing the vast amount of relevant research papers uploaded to Arxiv.org with related companies mentioned on Crunchbase.com, a business directory, to see which countries have put deep learning to good use the most. The authors used a neural network to sift through 1.3 million documents on Arxiv and incrementally narrowed down search parameters by analyzing titles, keywords, locations, etc., before comparing the list with Crunchbase registry analyzed the same way.

As expected, the deep learning papers covered topics of computer vision, computer learning, machine learning, AI and neural networks, with the US producing some 30% of all deep learning research papers and 30% of all other unrelated research papers. China was overrepresented in the deep learning section, producing three deep learning papers for every one unrelated to deep learning. Computer vision and computer learning were the most common topics, jointly encompassing some 70% of all deep learning papers on Arxiv. California was the most common location mentioned on Crunchbase, with 15% of all listed companies having headquarters

[41] https://arxiv.org/pdf/1808.06355.pdf

there. Texas ranked highly as well, which is explainable by the fact most disillusioned Californians named the barbecue state as the most likely US resettlement destination in the 2018 Bay Area Council Poll[42].

The analysis showed that China has the fastest rise in deep learning-related business ideas, with European countries falling behind and France being the very worst. The explanation for this effect is that Chinese business, research, and manufacturing sectors exist in tightly clustered regions, with the Chinese government having lax regulations on any research that promotes business growth and advances Chinese supremacy on the global market of ideas. This kind of industriousness does tend to produce items of subpar quality but fosters innovation, cost cutting, and quick turnaround.

The paper goes on to compare deep learning to seminal inventions such as the steam engine, electricity and the free exchange of information known as the internet, noting that each of these led to the rise of an empire: the UK conquered half the known world thanks to the steam engine, the US boomed because of electricity and Silicon Valley would command nothing without the internet. This would make us believe that the research in deep learning and AI-related business technologies will boost China to the position of a world superpower; any country that *doesn't* have an economic strategy focused on deep learning and closely mimicking China's is bound to fall behind.

Rise of a new empire

For an invention to be of such earth-shattering magnitude as electricity, it should have three distinct qualities: rapid growth, diffusion into new areas, and a high degree of impact in those fields. Neural networks that use deep learning practically teach and grow themselves, with the added benefit of their owners being able to pit

---

[42] https://www.scribd.com/document/380910605/2018-Bay-Area-Council-Poll-More-Plan-to-Exit-Bay-Area#from_embed

them against one another and see what comes up. We also find that deep learning is getting more and more practical applications with each passing day as some long-standing problems in various industries that were simply too costly to do any other way are now being reexamined. Finally, neural networks and deep learning can provide genuinely novel insights where applied and are able to boost productivity beyond human capabilities. The authors of the paper also answered in the affirmative to all three but by using a neural network to analyze publication dates, diversity of topics mentioned, and references used before comparing numbers on a year-by-year basis.

The steam engine was what started the industrial revolution in 19th century UK, with raw muscle strength being replaced by steam pressure, but it was electricity that helped in miniaturizing every aspect of factories in the 20th century and the internet that provided an instantaneous information flow that will make deep learning a transformative force for the 21st century. Each new quantum leap was always marked by the discovery of brand-new materials and ways to extract even more resources from old ones for less cost. If we now look back at how long it took for an industry to adopt one such revolutionary invention, we can note the industrial giants were set in their ways and incapable of adapting for several decades; it was always the small, nimble competition that seized the **first-mover advantage** in an environment of legal uncertainty that allowed unbridled experimentation.

In a well-set industry, such as cellulose production through pulp-and-paper mills, producers have razor-thin profit margins due to just how much legislation there is, with governments regularly adding even more. For example, some cellulose waste chemicals are known to be toxic if released into the water, but for others, there's only suspicion and no concrete proof; a pulp-and-paper mill owner would be tempted to use new and cheaper but potentially toxic chemicals as much as possible before the government outlaws them. Once out in the wild, these chemicals have unknown effects on plant life and animal health, which is even worse than if they were poisons – since poisons have known effects and treatments. There is no clear solution to this dilemma since we do need paper but can't help polluting when producing it. This leads us to **tragedy of commons**, the unavoidable outcome of such business mindset on the environment and resources we share.

The air we breathe, the water we drink and the very soil we live on are considered joint properties, a shared resource we all need and compete for but can't really affect them much; it's the businesses set to exploit as much as possible before anyone else does that pollute and ravage the environment in their mindless pursuit of profits. In 2010, "Deep Horizon", a Gulf of Mexico oil rig owned by BP Exploration & Production, exploded killing 11 workers and releasing 4 million barrels of oil into the ocean over 87 days until the leaking oil well was plugged. BP Exploration & Production was eventually sued by a party of litigants claiming damages and had to pay out over $20bn on top of US government issuing a $5.5bn penalty for water pollution and $8.8bn for damaging natural resources[43]. So what caused the Deep Horizon catastrophe?

Executives at any company have two guiding principles: **duty of care** and **duty of value**. The former mandates that they do their due diligence before undertaking any projects to make sure their company doesn't harm the environment and to constructively contribute to a better society for everyone. All these notions of care are idealistic, but more importantly, *they're impossible to quantify*. On the other hand, we have the latter principle that states an executive must do whatever it takes to increase company value, determined by comparing revenue numbers, *which are quantifiable*. These two principles are meant to balance out, but that's never the case and all companies that survive for decades gradually become more and more exploitative, ruthless and manipulative to squeeze

---

[43] https://www.epa.gov/enforcement/deepwater-horizon-bp-gulf-mexico-oil-spill

out that extra 0.1% profit that makes the next quarterly balance sheet green and provides the CEO with a fat bonus.

If we now look back at the consumer market over the decades, we'll easily notice companies, such as telecom operators, that experienced this inevitable transformation and became monsters that overcharge and ignore customers to the point their contractual obligations border on a scam. It's not that executives gleefully enjoy causing distress but simply that any company that wants to survive has to increasingly strangle existing revenue streams without investing anything more or adding any value to the customers. Even Google learned that lesson as they dropped the "Don't Be Evil" motto; moral values are antithetical to profits, and companies that want to make money must be willing to consider treading the line between good and evil, if not outright dashing over it before someone else does.

In a fresh market, customers would flock to a competitor, but in a highly-regulated market, that company has a monopoly, and *there are no alternatives*. We only have to look at Facebook's acquisitions to see how this plays out. In 2012, Facebook bought Instagram[44], a popular image-sharing social network, for $1bn and thus got its technology, brand, user base and all the users' private data; even when someone does make a viable alternative, the tech giants swoop in and devour the competition just like a cheetah does to an antelope. For Instagram creators, this was a dream come true, and they're set for life, but for Instagram users, it's back to the old crummy Facebook paddock.

The thing is that the *internet is another one of those shared resources*, with the only difference being that it's not physical, but we sorely need it nonetheless. However, there is barely any legislation that would prevent pollution of the internet or enshrine the rights of the ordinary users when it comes to sharing information and original content. Just like with soil, water and air before the

---

[44] https://dealbook.nytimes.com/2012/04/09/facebook-buys-instagram-for-1-billion/

industrial revolution, any company can do as it pleases online with abandon. This is the future of the technology market, and with the arrival of deep learning, things will only get worse as a company headquartered in India, China or Texas can create a worldwide product powered by deep learning; when the product starts falling apart at the seams, due to being pushed beyond its theoretical limits in pursuit of profits, the victims will have absolutely no recourse.

Deep Horizon disaster thus shows an inevitable conclusion of a profit-driven corporate system where environmental catastrophes must necessarily happen because the reward is simply too great to ignore. Naturally, the corporations fostering the development and deployment of smart machines into the general public don't care about any long-term consequences; why would they? Whatever brings the largest profit *right now* while barely remaining within the legal bounds is all that matters, and once that had become the corporate norm it's only a matter of time before the entire humanity gets to foot the bill of wanton smart-machines research.

This would also imply that BP Exploration & Production actually got off cheap as they most likely caused thousands of oil leaks in their reckless drilling expeditions – it's just that the Deep Horizon leak was too big to ignore and they paid the price. When this logic is applied to deep learning and neural networks we get a bleak picture of rampant experimentation that populates our digital environment with all sorts of shabbily made assistants, such as Google Translate, probably the most famous translator.

Google's babbling translator

Unimaginatively dubbed **TranslateGate**[45], the controversy around Google Translate arose when someone typed in the word "dog" 22 times with a single space between each word and translated it from Maori to English. The resulting text was, "Doomsday Clock is three minutes at twelve We are experiencing characters and a dramatic

---

[45] https://www.rt.com/news/434055-google-translate-dog-apocalypse/

developments in the world, which indicate that we are increasingly approaching the end times and Jesus' return" (sic). Granted there is at least one other case where the exact same word repeated over and over in writing produces coherent sentences without including neural network translation – The Lion-Eating Poet[46].

The origin of The Lion-Eating Poet story is a 19th-century Chinese linguist trying to demonstrate just how difficult it is to use Chinese when written using the Latin alphabet, which is called "pinyin" and was advocated as an improvement over traditional Chinese pictograms. The entire story consists of the word "shi" written out 98 times and talks about a poet named Shi who lived in a stone room and decided to eat ten lions, buying them at the market and trying to eat their meat; to a native speaker of Chinese the spoken story is perfectly understandable due to various pronunciations and inflections, but it's a mess written out in pinyin. This example emphasizes just how much of language we use is contextual and that complex languages, Chinese in particular, seem to be losing a lot of nuance with the adoption of foreign words and writing customs.

In any case, TranslateGate revealed that typing in "bi ng is be tt er" (Bing is better) and translating it from Somali to English produces "it is up to you" and "üüüüüüüüüüüüüüüü ääääääääää ööööööööööö" from Estonian to English produces "Nightly work-outs for the most part of the project". There are many other such examples at the TranslateGate subreddit[47], but the most notable part is that all of them consist of translating from internet-obscure languages into English, which is considered the universal language of the internet; it's quite likely there are many other similar language surfaces where the translation is slightly or entirely off, but we'll probably never find out the full extent of just how wrong Google Translate is.

---

[46] https://www.yellowbridge.com/onlinelit/stonelion.php

[47] https://www.reddit.com/r/TranslateGate/

For now, Google Translate sort-of works to the point Google includes it in their product suite alongside Gmail and Google Docs, meaning they're happy with the performance. One more detail with Google Translate is the option for any visitor to input an alternate translation, helping the neural network – over time, volunteers might enhance the accuracy and contextual acumen of Google Translate, but it's unlikely ever to be a proper standalone product that can carry its weight in a real-world environment. If Google can't do it, who can?

Shenzhen, the powerhouse of China

Located just north of Hong Kong, the fifth busiest port on the planet is the small market town of Shenzhen. During the 1990s, Shenzhen experienced tremendous growth, becoming a sprawling center of Chinese research, development, and manufacturing that now covers 750 square miles and includes literally hundreds of factories. Got a cool idea? Hong Kong has smart businesspeople who are willing to listen and have the means to order a prototype from a Shenzhen factory by the end of the day. If it works, millions of copies can be made by the end of the week, carted to Hong Kong and exported to the entire world.

By being joined at the hip, Hong Kong and Shenzhen cover all the basics and comprise a powerhouse of China that's been granted special exemptions from Chinese government regulation and taxes. There is nothing quite like it in the world, and unless other countries, in particular, the English-speaking ones, step up to the plate, they'll be outnumbered and outgunned. There is one downside of such explosive growth and blazing turnaround – the notion of lower quality products and services churned out by the millions with little consideration for standards, at times literally lighting a fire under our feet.

In 2015, hoverboards were the coolest fad, with everyone of any import gliding on one. Shenzhen alone had 300 factories churning out hoverboards 24/7, amounting to over a million units in just

October 2015, but their Li-ion battery packs were incompatible with voltages around the world, resulting in overnight fires and explosions; let's just say customers didn't warm up to the idea of Chinese hoverboards. UK, US, and even Chinese customers reported hoverboards going up in flames, and when the UK National Trading Standards put 15,000 Shenzhen hoverboards to the test, over 90% of them had a subpar electrical system or battery.

Fire departments across the world declared hoverboards a fire hazard. The media could barely contain the glee when showing the most dramatic hoverboard explosion footage, and retailers were suddenly stuck with thousands of hoverboards that couldn't be moved. Ironically, Shenzhen factories were stuck with warehouses full of working hoverboards that couldn't be sold simply because of a bad reputation and lack of electrical standards, but matters weren't helped by the fact Chinese factories generally engage in cutthroat business practices to lock out competition; this time around, they were all locked out of the international market due to the lack of an overall manufacturing strategy.

In March 2016, Chinese hoverboard manufacturers banded together to create a Hoverboard Industry Alliance that standardized manufacturing practices in accordance with US and UK electrical standards and asked for a set of hoverboard battery manufacturing regulations, which they got in May 2016 by UL, a safety company based in the US. The lesson learned here is that Chinese manufacturers do exhibit a casual indifference when making products aimed at international markets, but they're willing to make an about-face, cooperate and hold themselves to a higher standard when profits are threatened, making them a competitor to be reckoned with. This also implies we need a robust set of legislative limitations on deep learning, neural networks, and AI before the businesses start clamoring for it.

Normally the way regulations work is quite slow – new technology is introduced, and the legality or manner of its use is uncertain, there is some damage or death, the public demands someone thinks of the children and the government comes in with its habitual heavy-handed attitude. Investigations are started, years pass, committees are formed, and laws are made over the course of decades; it's simply how things have to work to maintain the integrity of the legal system.

It took the automotive industry years to come to grips with the fact seatbelts do save life and limb. Despite numbers being unequivocally clear on the matter, the car manufacturers fought tooth and nail not to have seatbelts while people died. This timeframe can't be applied to the development of AI since it will evolve on its own and jeopardize everyone while legislators twiddle their thumbs. If a significant portion of humans decide along the lines of "if we can't beat them, we'll join them" and implant an electronic device into their brain to have a direct uplink with the AI, the vanilla humans might end up just like chimps, causing a further divide between the rich and the poor.

Coping with AI advancement

Chimpanzees are uncannily similar to humans in that they too have a highly sophisticated social hierarchy, with superior males having the first choice of mates, sleeping locations, food, and drink; subordinate members pay great attention to avoid angering or provoking the leaders of the group by averting their gaze or slumping their shoulders to appear small. Chimps also chase status and compete with one another like we do, with males experiencing higher levels of stress that cause atherosclerosis and high blood pressure. Female chimps will groom one another, with the subordinate female caressing and fixing the hair of a superior female just like a human hairdresser would do with a client.

Chimps also go to war, and they're absolutely brutal at it – when another group of chimps intrudes on *our* territory, males go berserk and viciously attack the aliens with sticks and stones until they're gone. We've improved tremendously on the weapons of the chimps, but the essentials are remarkably similar, and we too are capable of exhibiting fervent hatred towards anyone wanting to take our tree stump. All these traits of status-hunting, supplication, tribalism, and aggression stem from a primal part of the brain humans and chimps share, the **limbic system**.

Found at the top of the spine, the limbic system is what houses all our deepest yearnings, drives and imaginations; since the limbic system is instinctual, it often reacts before we have the time to think, but we do create an elaborate story justifying its actions later on with our **neocortex**, the outer layer of the brain the limbic system is swaddled in. Neocortex was arguably a product of evolution that made us learn how to live and work together, how to suppress our instincts and behave like adults, for example by suing someone rather than trying to claw their eyes out if they slight us.

While we can consider chimps an oddity and wonder at how much they resemble us, there's no way to communicate with them or convey our thoughts – human ideas and concepts are so far beyond what chimps are accustomed to that they have no hope in ever understanding us. We also don't fear chimps or think of them as a threat but just let them be in their own cozy groves and meadows – that is until we need those areas for resources and just drive them off. What are they going to do, fight back? Our superior technology is more than enough to keep them completely under control: tanks, planes, rockets and nuclear bombs can deal with any chimp uprising in short order. Other than that, we can pretend we're friends as long as they know their place or just ignore them. When was the last time humans involved chimps in their decisions?

In case some humans decide to join their brain power with that of an AI and become **cyborgs**, humans enhanced with electronic implants, all other humans might end up just like chimps, living on such a primitive level that they're unable to communicate with these augmented humans. The cyborgs would essentially have superpowers and could be present everywhere, doing unimaginable things, which is close to what Oliver Curry, a British evolutionary psychologist, said back in 2007[48].

According to Oliver, the influence of technology will eventually lead to a rift between two parts of the humanity – tall, handsome and smart nobles and the short, ugly and dumb worker class. This process would occur over thousands of years and would essentially doom a large part of humanity to a genetic dead-end. If this sounds like H.G. Wells' *The Time Machine*, it's because that's exactly it. The novel was published in 1895 and tells a story of a scientist who invents a time machine and launches himself far into the future to witness the entire human race splitting in two: beautiful, carefree Eloi living on the surface and ugly Morlocks living in caves.

Oliver reckons that around year 3,000 the human race will experience its peak before starting a decline due to overuse of technology that would essentially *make us its house pets*, feeding on our emotions and brain impulses to power itself. Lifespan will increase to over 120 years, diseases will all but vanish, physical appearances will become stunning to signify excellent health, and racial mixing will eliminate different skin colors, producing a blend of coffee-colored people. Between then and year 100,000 is when all humans would lose all semblance of social skills thanks to technology eliminating the need to communicate face to face. Everyone would be cooped up in their own little chamber with the latest virtual reality implant providing them with all the stimulation they could ever want; compared to that, everyday life would seem utterly drab.

In a sense this differentiation is already happening as we speak, with those rich enough to buy the latest and greatest smartphones having unparalleled access to information databases and computing power our brains can't even fathom; think of this the next time you see someone walking down the street with their face buried in the

---

[48] http://news.bbc.co.uk/2/hi/uk_news/6057734.stm

smartphone screen, oblivious to the world around them and wrapped up in a bubble of their own. One interesting fact is that smartphones, computers and other high technology are highly addictive as they provide more stimulus than we'd ever encounter naturally – there doesn't seem to be an upper limit to this kind of addiction, and there's never enough computing power in the lives of such addicts, but it's their human organs that stop them from drawing too much.

Elon Musk's venture project **Neuralink**[49] tries to solve the problem of organic information bottlenecks, such as our natural vision being limited, by creating a "brain lace" that would be inserted directly into the brain stem, letting it tap directly into the internet or vice versa. This is extremely experimental and would likely result in total discombobulation of the hapless patient, meaning there are plenty of people eager to prostrate themselves under the scalpel. Neuralink would thus possibly become a third brain layer, enveloping both the limbic system and the neocortex but would serve exclusively to connect us with the godlike fount of digital power.

What's interesting is that certain brain diseases can be ameliorated using fairly simple electrical brain implants that release weak charges, but the complexity of a brain-AI interface is for now only the domain of science fiction; it's literally brain surgery, and no doctor wants to shoulder such liability. We don't really need to crack open skulls and implant neural interfaces in brains to achieve health care advancements using neural networks though.

---

[49]   https://www.theverge.com/2017/3/27/15077864/elon-musk-neuralink-brain-computer-interface-ai-cyborgs

Chapter 8 – Medicine with the Help of a Digital Genie

Neural networks promise to improve the medical field like no other invention. Health care is swamped with menial tasks that have to be done with absolute urgency and precision, such as drawing blood to check for blood sugar levels or measuring blood pressure, requiring an army of support staff that also has to maintain hygiene and do paperwork. This creates massive overhead and liability for any medical establishment; a doctor that has no clue about patient's blood sugar or blood pressure risks not reacting in time, prescribing wrong medication or a bad dose, resulting in patient's death or even worse, a malpractice lawsuit.

Even the lowliest medical staff needs extensive training to avoid injuring people, but a neural network can train itself by studying data or just creating a synthetic patient in its mind and practicing on it. Including a neural network in this procession of medical procedures would mean having a reliable, objective, tireless assistant that would also be capable of providing a second opinion on any kind of problem over the internet to reduce the legal burden of doctors across the globe, allowing them to act quickly and with certainty.

This lack of medically trained support staff is a huge problem in 3rd world countries where crucial medicine, such as dentistry, is still done by barbers using pliers. X-rays, in particular, take qualified staff to analyze, with the patient's life often depending on proper analysis. In cases where the staff isn't available, a neural network will do the job just as well or even better than a qualified technician over the internet or locally, paying for itself within a year and being able to work 24/7/365. However, there's no way neural networks and wearables will ever replace doctors, nurses or other assorted medical staff; wherever technology has been introduced in a sustainable and organized manner, it's always led to an *increase* in employment and productivity. The kind of future where machines and humans work side by side for everyone's benefit thus seems not only possible but the brightest one imaginable.

A neural network employed in health care could have instant access to aggregate results of millions of cases on any given health problem to spot symptoms and predict disease progression, which is something doctors are still woefully unprepared for. Such helper could make use of high definition scans in a way no human could, zooming in and scanning each pixel for signs of tissue change, comparing the results to typical disease progression and suggesting medication; when it comes to medicine doctors need all the help they can get.

The human body is wonderfully weird and takes a well-trained team of medical experts even to begin figuring out what went wrong when disease strikes. We first notice faint aching and some discomfort before the pain proper reaches unbearable levels. That's when we're no longer able to function normally and have to face doctor's judgment. The interesting part is that the outward symptoms can be barely noticeable, so deciding to consult a physician with early signs of disease might get us the squinty look unless we insist on getting lab tests done that would confirm our daily experience.

Separate organs within the body act on one another and are impacted by the outside environment in real time, making the disease ebb and flow. This is particularly obvious with chronic diseases, such as diabetes, the disruption of pancreas and liver function that leads to uncontrollable blood sugar levels that cause cells to combust due to energy overload. The 2017 National Diabetes Statistics Report concluded that 10% of the US population has diabetes, making it the 7th most common cause of premature death in the US. Diabetes truly is an epidemic that affects everyone from six-month-old babies to adolescents, and part of the reason why Affordable Care Act (ACA) was pushed through with such haste – minorities tend to be susceptible to diabetes and employers avoid hiring what they see as a liability in terms of health care costs.

The elderly are especially vulnerable to diabetes as it impacts the brain, disrupting its delicate blood sugar regulation to cause dementia and Alzheimer's, knotting of brain cells, to the point doctors are already vying to call it "type 3 diabetes". There is no cure for diabetes, but it can be managed through healthy lifestyle choices; proper treatment should involve an entire community helping the diabetic take medication when needed and keep track of blood sugar levels. The diabetic can't be counted on to take care of themselves, so there needs to be an entire support network supplying them with healthy food and urging them to stay active, on top of medical professionals checking up on the diabetic.

Wildly oscillating blood sugar caused by diabetes can also incite mood swings, hallucinations and fits of rage. In youthful people, these side effects are painful, but in elders who live on their own and don't know how to seek health care, they're just tragic; once they experience loss of muscle mass (sarcopenia) and loss of bone density they are in danger of having a bone-shattering fall that causes complete immobilization and thorough dependency on outside help. If the diabetic has also happened to drive away people who care for them the most, the only alternative is government health care, requiring massive amounts of money just to keep this one person in a wretched state of subsistence.

Diabetics cause a massive strain on any health care system since they on average spend about 230% of what a non-diabetic does, with 2012 stats showing the total cost of health care for US diabetics at $245bn a year. With life expectancy rising, it appears the future will be full of moody, frail and immobile elders who have no idea where are they are and how they got there, with the rest of society fully employed caring for them. That is unless we can make neural networks that will help out.

By the time we notice disease symptoms, it's already well underway, but medical implants and wearables could provide a constant stream of body values to a neural network that could estimate risks of disease, allowing for constant low-key monitoring of values such as blood glucose and blood pressure at a very low cost. It's already happening on the consumer market and Apple Watch 4, revealed September 2018, is set to ship with the ability to monitor pulse and blood pressure for signs of a heart attack[50]. The market of slick medical wearables is a wholly untapped one, and Apple is doing a smart thing catering to the rich and sensitive to test out the technology that also doubles as a status symbol.

Even those who can't afford Apple products can benefit from medical wearables worn at home that report findings to the doctor. A 2009 Canadian study[51] examined 26 studies covering 5,069 diabetics and found those using such home telemonitoring technology had a much better quality of life compared to those who simply had access to telephone support. Overall home telemonitoring helped diabetics keep their blood sugar in check, reduced the number of incidents requiring hospitalization, and lowered the number of days a diabetic was hospitalized.

Diabetes impacts every system in the body, throwing internal self-regulation mechanisms in disarray, but the sorest bodily points are feet, skin, and eyes due to fragile vascular systems present there that get hammered by high blood sugar and high blood pressure, so it's no surprise the majority of wearable implants focus on ameliorating the impact of diabetes in those three. The most common way to use

[50]   https://www.zdnet.com/google-amp/article/apple-watch-4-why-digital-healths-future-depends-on-apple-finding-a-partner/

[51]

https://www.researchgate.net/publication/26296275_Home_telehealth_for_diabetes_manag ement_A_systematic_review_and_meta-analysis these wearables and implants right now is by having them communicate with a smartphone app using binary states – the problem is or isn't present but neural networks would be able to gather data on a continuous basis to create a finely tuned diagnosis that perfectly fits the patient and helps them recover in the most effective way possible.

FreeStyle Libre skin patch by Abbott pharmaceutical company is worn on the back of the arm and measures blood sugar using a minute filament that scans the arterial blood flow. The skin patch reports to a handheld scanner that works through clothing, so the patient doesn't even have to get undressed, but these wearables can just as easily connect to a smartphone app to report status to the diabetic. Eccrine Systems, Inc. made a skin patch that samples sweat for blood sugar readings and can release insulin back into the bloodstream without using needles – there we go, no more pricking and bloodletting. GoogleX is Google's experimental branch in charge of all sorts of wacky projects and proof-of-concept gadgets, such as contact lenses that sample tear fluid to measure glucose while helping the diabetic see better. The idea was patented in 2015.

Finally, diabetics can be helped with smart socks to keep diabetes under the heel. Tingling and poor peripheral circulation in feet are first signs of impending diabetes with the exact cause again being high blood sugar and blood pressure. If left untreated either or both feet or even legs have to get amputated to stop the rot from spreading to the abdomen; socks with embedded heat sensors would be capable of detecting blood flow under the skin and warning the person through a smartphone app to start moving to kickstart the circulation or stop standing on a single foot while helping find small cuts that get infected in diabetics at an alarming rate.

It sounds laughable, but some of the smartest minds in tech are working on smart socks for diabetics since costs of merely keeping them alive are so immense that a mere 1% reduction would be worthy of a Nobel prize. Imagine German PhDs from the Fraunhofer Institute and US PhDs from the University of Arizona huddling around to create a smart sock with the most advanced sensors on the market and getting foiled – by a washing machine. So far, no smart sock design has survived repeated washing, but Smart Sox should still be in consumers' sock drawers by 2021.

Why stop at wearables? Some tech is safe enough to be implanted right into the body under local anesthesia without hospitalization that would take up a precious bed – just release the diabetic home, check up on them over the internet and drip the drug remotely; they don't even have to lift a finger. Right now, diabetes drugs come either as pumps or injections, both depending on the diabetic to administer the right dose at the right time, but imagine someone in that condition going on a vacation and carting around insulin kits – it's enough to send their stress levels through the roof. The first idea behind implantable tech for diabetics is an artificial pancreas.

Viacyte is preparing VC-01, an artificial pancreas that is still going through testing on four brave diabetic volunteers who are sick and tired of administering insulin; it contains stem cells and should arrive on the market in 2021. Joan Taylor of De Montfort University of Leicester is a UK professor researching minimally invasive medical implants. In this case, she came up with a watch-sized pancreas consisting of biologically compatible gel that gradually releases insulin based on blood glucose levels. Intarcia is developing ITCA 650, an insulin implant the size of a matchstick that will drip exenatide, a helper drug commonly taken with metformin, the primary drug for treating diabetes symptoms. The usual way to administer exenatide is by poking the abdomen one-two times a week, but ITCA 650 will last for up to a year before needing a replacement.

Should we trust this kind of technology? All technology has a small chance of catastrophic failure, with electronics being particularly vulnerable to electrical discharges from the sun (aka solar flares) and other magnetic disturbances. This applies to medical wearables and implants too, but the idea is always to have redundant systems ready to kick in if electronics fail. It's a good idea to back everything up in paper or offline form for a "just in case"; storage is dirt cheap and makes any tech meltdown a nuisance rather than a disaster. Software used in all computers has bugs and although this doesn't technically apply to neural networks they can still experience the human equivalent of mental breakdowns and start producing gibberish results due to unknown causes, like we saw with TranslateGate.

It's up to each individual to ponder the question of smart technology, such as socks and implantable devices, and decide if that's what they really need on and in their body. Only if the pros outweigh the cons should such potentially intrusive technology be a part of our existence; we should neither outright embrace nor dismiss anything that could fundamentally change our lives. In the meantime, the research on how to implement neural networks in health care rolls on.

Chest X-rays

A paper titled "Can Artificial Intelligence Reliably Report Chest X-Rays?"[52] examines the case of a neural network trained using 1.2 million X-ray images and the deep learning process to assist radiologists where there are staff shortages or where the staff is inexperienced. The neural network was specifically trained to detect nine chest abnormalities and later tested on a sample of 2,000 unused X-rays against the majority of three human radiologists, showing performance comparable to humans. The idea is that low resource areas of the world have greater access to X-ray machines than personnel capable of reading them correctly, so such a neural network could access remote X-ray images of patients and provide an instant diagnosis in cases such as tuberculosis, where an X-ray can provide more information on disease progress than any other clinical test.

---

[52] https://arxiv.org/pdf/1807.07455.pdf

Lung disease estimation

"Deep Learning from Label Proportions for Emphysema Quantification"[53] shows how a group of Danish doctors trained a neural network using deep learning to identify the severity of emphysema (abnormal enlargement of alveolar tissue) in patients. The neural network was first taught using samples labeled by human doctors, such as "1-5% emphysema" and then independently tested on samples labeled using traditional methods, such as lung densitometry. The neural network outperformed all known methods for assessing emphysema by 7-15% thanks to a hidden architecture layer that estimated emphysema volume based off of labeled samples provided by humans and could correctly predict the spread of the disease on par with human specialists.

"Quantification of Lung Abnormalities in Cystic Fibrosis using Deep Networks"[54] is a collaborative research paper by Dutch, Portuguese and Danish doctors trying to salvage whatever lung tissue is viable in patients struck by cystic fibrosis (CF), a genetic disorder that most commonly affects Caucasians. CF materializes as abnormally thick mucus secreted in the lungs that clogs the fine alveolar capillaries, causing infections and trouble breathing; mucus is normally slick and freely flows to prevent the mucous membranes from drying out. Drugs do exist to treat CF, but they're given on a daily basis using vein injections that can damage veins and cause a catastrophic loss of self-confidence. Since CF is genetic, there's no cure, but early treatment can help tremendously in saving the patient's lungs and thus their quality of life.

Two layers were put to the task in the neural network – the first one labeled diseased and healthy tissue while the second classified three

---

[53] https://arxiv.org/pdf/1807.08601.pdf

[54] https://arxiv.org/pdf/1803.07991.pdf

different problems on damaged tissue: damage to the airways, mucus blockages, and deflation of alveoli, grape-like clusters that absorb oxygen into the blood. Different layers were necessary to finely tune disease detection that would otherwise be too biased towards false positives. By using 194 heatmap pixel-precise CT scans of CF children that were on average nine years old, the neural network was taught how to recognize the early symptoms, progress and the most likely path of development. Out of 194 patients, 50 didn't show any signs of the disease, so they were used as a test.

A fine grid was placed over the lung scans, with each square measuring roughly ½ by ½ inch. The neural network was taught to recognize different textures and was told to mark each square as diseased if more than 50% of its surface showed such texture. By carefully weaving two layers and thus weighing the neural network's learning patterns, the authors of the paper achieved nearly 50% greater accuracy in CF scan analysis than a single-layer neural network used in similar procedures.

Brain tumor estimation

Tumors too can be sized up using deep learning-trained neural networks, as shown by "3D Convolutional Neural Networks for Tumor Segmentation using Long-range 2D Context"[55]. Traditionally trained neural networks show amazing accuracy in assessing tumors from sequential MR (magnetic resonance) imaging but their computational cost prohibits the concept from scaling. However, neural networks trained with deep learning can overcome the architecture limitations to combine 2D images into a 3D voxel representation of the organ, in this case, brain and the parts affected by gliomas, the most common type of brain tumor. The network can even fill in the gaps where some of MR images are missing by its nodes voting on the most likely outcome and having it assess the votes. The biggest upside of this neural network is that it

---

[55] https://arxiv.org/pdf/1807.08599.pdf

standardizes work normally done by humans who differ greatly in their estimate of the affected areas.

Heart murmur detection

"Murmur Detection Using Parallel Recurrent & Convolutional Neural Networks"[56] proposes a novel way to detect heartbeat irregularities that can be heard through the chest using a listening device, which is why doctors wear stethoscopes. Heart murmurs occur either due to physiological causes such as stress or due to serious defects such as heart valve deformation. The latter is particularly grave and may indicate the person is suffering from health issues such as fever, anemia, thyroid problems and high blood pressure. The doctor determines this by listening to heart muscle activity that should consist of two strong thuds due to contractions and minor sounds in between caused by blood flow and valve activity. It takes years of experience to distinguish benign from malign causes of heart murmur, which is where neural networks come in.

The murmur detection paper suggests presenting the heartbeat of any given patient as a waveform, meaning a set of data points in two dimensions that can be visually and acoustically broken down into particular segments for fine-grain assessment. The acoustic data was gathered from open source medical datasets totaling 3,040 normal and 143 murmur recordings, processed a little bit to eliminate any noise, and fed to two distinct neural networks working together to distinguish one from the other; they ended up having 87-98% efficiency when it came to matching normal to normal and murmur to murmur data.

Sizing up prostate cancer

"Epithelium segmentation using deep learning in H&E-stained prostate specimens with immunohistochemistry as reference

---

[56] https://arxiv.org/pdf/1808.04411.pdf

standard"[57] suggests a way to help estimate the degree of prostate cancer using neural networks. Considered the most common form of cancer that only impacts men, prostate cancer has 1.1 million new worldwide diagnoses a year and the first sign of trouble is a high blood level of a certain antigen, tested with bloodwork or an enlarged prostate, which is tested manually.

The traditional way of confirming prostate cancer is by insertion of a hollow needle into the prostate to take a small tissue sample that is then stained using hematoxylin and eosin (H&E) that turn from blue to red depending on tissue composition, enabling easy recognition of cytoplasm, muscle, collagen etc., by a medical professional. Each such stained slide is graded to estimate the spread of prostate cancer and chances of remission. This is a tedious, time consuming but critical part of the diagnosis, with the prostate tissue sampling repeated up to a dozen times on a single patient to confirm previous findings.

The problem of automated prostate cancer detection is that a slide generated from a prostate tissue sample can have plenty of noisy data points, such as inflammatory components. Such slides would then have to be laboriously annotated by medical experts to help train the neural network, but the authors of the paper solve this problem by attempting image analysis on a pixel level. The training dataset used 102 digitized slides of prostate tissue samples, and the neural network achieved 89% precision, with the authors noting that staining slides should be of a higher resolution to allow proper zooming into for higher accuracy. The goal is to eventually have a fully automated prostate cancer detection algorithm that will outline potential cancer areas on the slides with 99-100% precision and the doctor confirming the diagnosis.

---

[57] https://arxiv.org/pdf/1808.05883.pdf

Alzheimer's disease prediction

We mentioned diabetes and how it impacts the brain, in particular, how it makes proteins accumulate in certain brain tissues to cause knots that lead to Alzheimer's disease (AD), turning a person into a shadow of its former self. Predicting not just AD but how any patient will turn out is a thankless task because each disease has a set of common symptoms and a kind of flourish that is unique to that person and has to do with their genes, environment, habits, etc. What doctors commonly do is look for these commonalities and treat them but ignore the flourish, meaning the disease has been subdued but not defeated.

This combination of common symptoms and a unique flourish is why doctors mostly give an approximate timeline of mental regression in AD cases, but neural networks might help us go into the gritty details, unravel the progression of this wicked disease and know the exact patch AD will take in any given case. The research paper "Using deep learning for comprehensive, personalized forecasting of Alzheimer's Disease progression"[58] looks at data from 1,908 patients with cognitive impairment or AD and tries to figure out how cognitive tests, laboratory results and clinical signs predict each patient's progression of their ailment.

A neural network is given all the patient data and is allowed to use unsupervised learning to predict how the tests, the results, and the signs correlate, allowing medical professionals to fast-forward to any future point in the state of an AD patient and see their condition. This simulation model is aptly termed "computational precision medicine" and involves heavily preprocessed medical data on each patient, such as cholesterol, potassium, hemoglobin and triglyceride levels in blood, weight, age, geographical region, and heart rate, giving predictions and assigning probabilities to each.

---

[58] https://arxiv.org/pdf/1807.03876.pdf

The conclusion was that the neural network correctly showed up to 18 months of dementia progress and performed on par with supervised training neural networks, helping medical professionals build detailed AD profiles and in-depth risk estimates. It's quite likely this kind of neural network approach will lead to the creation of personalized medicine where there are no unnecessary tests, anguished waiting and wasted medication as doctors can finally head this and other diseases off at the pass.

Synthetic patient generation

A modern hospital in an urban center has a large throughput of patients and a chronic trouble of recording all their symptoms. Medical staff might not pay attention to all the tidbits of data, doctors might not have the most readable handwriting, and even the patients themselves might not care about paperwork – we just want to get everything over with as soon as possible. In short, this creates a mess that compounds as patients keep pouring in and all the illnesses blend into one big blob. How do we digitize this data? How does anyone search or classify any of this?

"Synthetic patient generation"[59] presents the idea of using neural networks to create detailed patient profiles and all the symptoms associated with any given disease. The neural network was first fed accurate medical data spanning nine years of Tanzanian hospital operation and then trained to recognize underlying patterns. The data consisted of gender, age, symptoms, diagnosis, time of year, tests ordered, their results and treatment. The idea was to fill in the blanks in cases where patient's medical data is unknown or guesstimate the diagnosis using background profile data, not with absolute certainty but with a great degree of certainty.

The neural network used for this task consisted of an encoder that transformed all data inputs into lower-dimensional vectors that were then fed to the decoder and the output compared with the input, thus

---

[59] https://arxiv.org/ftp/arxiv/papers/1808/1808.06444.pdf

training and testing the network in one go. Raw and decoded data displayed together on a scatterplot shows how the neural network, at first, clustered all the decoded data, but then, eventually, learned to finely distribute it to the point results that were slightly off but otherwise indistinguishable from actual patient data. In the end, doctors were asked to evaluate a dataset of patient symptoms consisting of real and synthetic profiles, with the goal to distinguish which is which. Doctors identified 20% of synthetic profiles as synthetic, 23% of real profiles as synthetic and 80% of synthetic profiles as real.

Predicting medication effects

The pharmaceutical company has to invest a lot of time and money to push medication into an open market. Since each body reacts differently to the same substance, there is a huge inherent risk in trying to predict effects *and* side effects; it's a gamble that better pay off. Viagra (sildenafil) was actually tested as a hypertension medicine in the 1990s until the volunteers reported powerful erections so the pharmaceutical company marketed it as such to recoup original research costs that can go up to billions and not yield a worthwhile drug – Viagra netted $2bn in 2008 alone.

"Deep learning for in vitro prediction of pharmaceutical formulations"[60] aims at a lofty goal of training a neural network to analyze known medication side effects and infer what will happen when new medication is given to humans. Just like we saw in other examples, where a neural network is involved in crunching data, there is no absolute certainty but a degree of possibility that comes with a risk of error; cost-saving that comes with moving away from expensive clinical trials should be enough to offset the risk. Aided by classical computing to create drugs from scratch, clinicians can sift through thousands of different medications and immediately discard the most harmful ones while using accumulated reports on drugs that were tested to find the most promising substances.

---

[60] https://arxiv.org/ftp/arxiv/papers/1809/1809.02069.pdf

Neural networks were previously used in a similar vein to predict possible liver damage of drugs to good effect, better than any other machine learning model. Since this kind of data is generally not available for easy analysis, the neural network can fill in the blanks in existing data and estimate the future clinical trajectory of any new drug. In this case, the neural network was tested on predicting water solubility of drugs by using 276 known descriptions gathered from medical knowledge repositories; the testing dataset consisted of nine separate values for each drug that described its complexity, such as molecular weight and hydrogen bond counts. The neural network correctly predicted solubility for 95.57% of quick-dissolving oral pills and 82.02% for slow-dissolving ones.

Fetal ultrasound analysis

Knowing that she's carrying a clump of joy in her womb can be the most exciting experience for a woman and one that also fills her with trepidation and uncertainty. There are very little doctors can do to guarantee a baby's proper development except observe, which is usually done with ultrasound. Seeing their baby on a grainy, low-resolution screen is generally a solemn occasion for the happy couple, including the meek father who obligingly nods and smiles, but doctors actually use ultrasound to assess the development of baby's head, the most delicate indicator of any deformities. If the doctor's not sure, he will order tests, more tests, and even more tests but with the help of neural networks, this might be wholly unnecessary.

"Automatic evaluation of fetal head biometry from ultrasound images using machine learning"[61] aims to help doctors who, up to this point, had to manually measure and estimate the baby's head circumference, the most obvious sign of health. The artifacts mentioned above in ultrasound imaging are yet another problem, but neural networks have a solid dose of error tolerance and a way to

---

[61] https://arxiv.org/pdf/1808.06150.pdf

replace lost data with fuzzy guesstimates. Neural networks generally benefit from having a well-annotated dataset, meaning that they should be trained or let to train themselves using high-quality data but in the case of fetal ultrasound this kind of data is often missing or is incomplete.

The neural network is trained using what's known about fetal head anatomy to detect head boundary pixels in the image and then drawing an ellipse around the head. In cases where the doctor has set an ultrasound in an incorrect position, the neural network will use known rules of ultrasound propagation to differentiate maternal tissue, such as placenta, from the baby's head.

Training was done with 102 actual baby ultrasound images taken with Samsung WS80A ultrasound machine in Seoul; an additional 70 images were used for testing. The test was to correctly identify the baby's skull and parts of the brain: cavum septum pellucidum and ambient cistern. Two doctors were independently given the same test and afterward separately evaluated each other's and the work of the neural network. Doctors on average found that 87.14% of the neural network's work to be correct; for comparison, the doctors found each other's work to be 100% correct.

Autism detection

Dustin Hoffman did a phenomenal job portraying a young autist-savant that gets bequeathed millions in *Rain Man* to the point the movie garnered four Academy Awards in 1989, but life isn't as rose-tinted for actual autists; they're often unable to function on a core level without a caregiver. Despite autism having the potential to be a deeply debilitating disease, we still don't have the faintest clue what causes it, which is why autism is generally thought of as a "spectrum disorder", with some autists having mild impairment that can be aided by discipline and others having seriously disturbed brain chemistry. Earliest signs of autism are seen as soon as the child is capable of fine motoric movement as it often stacks or lines things up according to its own unfathomable rules, but there was no definite way to establish a diagnosis up to now. The emphasis is on "was" since now we've got neural networks.

"Brain Biomarker Interpretation in ASD Using Deep Learning and fMRI"[62] posits using neural networks as a way to peek into the brains of children with an autism spectrum disorder (ASD) and figure out what's going on. The first step in tackling ASD is to compare changes in blood flow in brains of normally functioning individuals and those with ASD using an imaging technique known as fMRI (functional MRI). 82 kids with ASD and 48 healthy ones underwent brain blood flow imaging as they followed a set of dots on the screen, with the images being compared by the neural network that found the difference in brain regions being activated. Overall ASD kids showed 50-100% more brain regions being activated compared to healthy individuals, implying that autism is when the brain is put into overdrive by an unknown set of causes.

Early cancer detection

Cancer is a vile disease, one that surreptitiously develops in the body and hijacks bodily systems one by one, modifying gene expression in cells in such a way that it becomes unstoppable. For example, cells have an expiry date and a gene for apoptosis, which is, in essence, a self-destruct button, but cancer disables apoptosis and makes the cell immortal. Cancer Research UK stats[63] from 2013-15 show that every two minutes there is a new cancer diagnosis in the UK alone, coming out to some 366,000 new cases every year and almost half of them diagnosed in late stages when treatment becomes a desperate struggle for survival. Doctors are really starting to lose the race against cancer, and all seems lost to the point they might as well start cheating by using neural networks for early detection.

---

[62] https://arxiv.org/pdf/1808.08296.pdf

[63] https://www.cancerresearchuk.org/health-professional/cancer-statistics/incidence#heading-Zero

"Adaptive Structural Learning of Deep Belief Network for Medical Examination Data and Its Knowledge Extraction by using C4.5"[64] suggests feeding comprehensive cancer patient data to a specially crafted neural network and letting it figure out when and where in the body cancer will manifest or has manifested to track down the cause. Since cancer patients do spend a lot of time in a hospital, there's an enormous wealth of medical data on the progress of cancer, but factors that lead to its appearance remain elusive. Even when controlled studies expose mice to these known carcinogenic factors, there's no telling how much that applies to humans since we move around freely and may be exposed to merely minute quantities of any factors, the combinations of which might lead to cancer proper.

FDA commissions a Report on Carcinogens every few years, so in May 2018, the U.S. Department of Health and Human Services drafted a report on Helicobacter pylori[65], a resilient bacteria that invade the digestive tract of humans and eats the protective lining, causing heartburn, cramps, ulcers, and even stomach cancer. About 36% of the US population and 50% of the total world population has H. pylori without even realizing due to poor sanitation and bad water quality[66], which is a huge issue in secluded areas of the nation and with minorities, but it's been found even in tap and bottled water in Sweden, with some research suggesting it can survive chlorination. There is some dispute as to how much H. pylori contributes to any given cancer type, but two researchers actually won a Nobel Prize in 2005 for definitively linking this nasty invader to gastric ulcers and indirectly to cancer.

Anyway, the authors of this paper decided to go with the black box approach to building a suitable neural network, with the idea that

---

[64] https://arxiv.org/pdf/1808.08777.pdf

[65] https://ntp.niehs.nih.gov/ntp/roc/draftmono/hpyloridraftmonograph_508.pdf

[66] https://ntp.niehs.nih.gov/pubhealth/roc/listings/hpylori/index.html

carcinogenic factors can be represented as functions and adequately approximated in a way nobody really understands but that produces valid results with a high degree of certainty. Patients' medical data, such as BMI, height, eyesight health, and hearing frequency was split into categories and represented as float, integer or code for the purposes of simplification for a total of 100,000 persons and their 5,900,000 pieces of data gathered during routine health checkups in Hiroshima, Japan, over the course of three years. Each person also had four images made – lung, breast and stomach X-ray and lung CT, with human doctors assessing each image and labeling it "normal" or "abnormal". The data was randomly split into 80,000 records used for training and 20,000 used for testing.

Testing showed that the neural network, named "Adaptive Deep Belief Network", accurately recognized which patients will develop lung cancer in 95.5% of cases and in 94.3% of stomach cancer cases. Fine tuning the neural network using two algorithms increased the accuracy even more – 98.1% lung and 98% stomach cancer detection rate. The two algorithms essentially worked as traditional software patches by repairing certain neural circuitry the neural network deactivated, thinking useless, and without delving into the black box architecture itself.

Final conclusions were that the neural network discovered white blood cell abnormalities as the most prevalent signal of cancer in general; abnormal levels of liver enzymes GOT, GPT and gamma-GPT as signs of stomach cancer; and albumin and total levels of protein as signs of lung cancer. Though these bloodwork levels aren't commonly associated with cancer, the interesting coincidence is that doctors have reported these abnormalities in cancer patients prior to the neural network's research. Since doctors are generally cautious in jumping to conclusions, these findings will be further corroborated by medical examinations before being included in the arsenal of tools for fighting against cancer.

Conclusion

Deep learning and neural networks promise an interesting future indeed. The conventional way we do science turned out to be grossly inaccurate to the point we get wildly different answers by simply changing the way we round numbers, but the scientific consensus so far was to dismiss these irregularities as statistical errors rather than demand a paradigm change. Scientists are now focusing on creating artificial intelligence, a thinking machine that can do the math for us and present its findings in a completely objective way devoid of personal bias. Still, things might go awfully wrong but not in a cliché *Terminator* way.

One potential future we could face with neural networks is that of narrowly defined and trained ones being hastily put to uses they're not meant for to capitalize on the feelings of euphoria, creating a slew of lobotomized digital assistants that are billed as brilliant. At least in the *Terminator* future the victims would be granted the sweet release of death, but in our timeline, we might suffer the tyranny of overpriced hardware sold as revolutionary just because it has access to a neural network that doesn't quite work as advertised; we'll pretend it does because it's the cool new thing to have.

The ultimate insight from fiddling with neural networks and artificial intelligence is that humans are fallible and that's okay. We can't ever reach perfection except in theory – meaning that any concrete action we do for that purpose, such as trying to create a perfect digital assistant, might cause us undue stress. The real source of danger would thus come from corporations poised to make as much money off of the neural network craze while endangering the very essence of our digital existence.

Though we've made structures that serve to soften the blow of evolutionary pressure on humans, the corporate business world is as cutthroat as it gets, rekindling the killing instinct in those corporate executives who can't start their day without a hostile takeover. Combined with artificial intelligence being a highly malleable, abstract technology we might reach a future where the drive for earnings outpaces all legislative, societal, moral and religious impositions to create an environment with obscene amounts of digital pollution, a World Wide Web filled with artificial stupidity. To paraphrase Aldous Huxley, "If you want a picture of the future, imagine a fist smashing on a keyboard forever."

Glossary

Addiction – An unnatural, non-essential urge to do something that ultimately harms the person. All modern business models center on creating addiction in users.

Algorithm – Sequence of commands written out for the software to follow. Any error in the algorithm inevitably produces bugs or glitches during execution.

Artificial intelligence – Capability of the digital brain to perform tasks on its own. So far only **narrow AI** exists (compare to **general AI** and **super AI**). There is a real possibility narrow AI might be marketed as general AI.

Black box – Program or device where it is not known or knowable how it operates, the emphasis is on its results. A certain margin of error is accepted.

Copenhagen interpretation – A way to avoid thinking or speaking about paradoxes and implications of **quantum physics** on real life. Can be summed up as "don't think, just calculate". **Schrödinger's cat** is an example of a scientist breaking this vow of silence.

Cyberspace – Digital representation of the universe with almost no time, space or natural resource constraints. Allows for unchecked **evolution**.

Cyborgs – CYBernetic ORGanisms. The term denotes a human that's been merged with technology to the point there's no distinction between the two. See **Neuralink**.

Deep learning – **Black box** approach to programming that creates adaptable software capable of evolution. Happens in **cyberspace**. Synonymous with **machine learning** but sounds much cooler.

Double slit experiment – Scientific experiment that proved observing electrons makes them a particle, but they otherwise behave like a wave, implying human consciousness can change reality. The origin of **quantum physics**. Attempts to trick an electron into showing its true nature revealed **quantum entanglement**.

Duty of care – One of two guiding principles for CEOs that states they should exercise due care for the environment, fellow humans and the entire humankind. See **duty of value**.

Duty of value – One of two guiding principles for CEOs that states they should do whatever isn't strictly illegal to increase revenue. Meant to be balanced with **duty of care** but always overpowers it.

Evolution – Self-optimization of living beings for the most resourceful handling of the environment. Happens on a timescale of geological significance. **Deep learning** and **machine learning** are attempts to have software cheat their way straight to end evolutionary results.

First-mover advantage – The advantage of small competitors to outmaneuver the established giants. Usually hampered by a comprehensive legal mesh mandating huge capital investments to merely start the business. Very strong in areas with legal uncertainty.

Flow correlation – A way to de-anonymize **Tor** users.

General AI – Digital intelligence thought to be as smart as a human. Has not been created yet but posited to evolve to **super AI** almost instantly.

Generative adversarial networks – Training smart machines by having them fight and learn from one another. A twist on the concept is having the machine battle itself.

Higgs boson – Theoretical particle that somehow causes gravity. Also known as "God particle". See **Large Hadron Collider**.

Ig Nobel – Mock award ceremony for the most ridiculous scientific paper or invention, held by actual Nobel laureates. Has a lighthearted undertone to it. A play on the word "ignoble": not noble in quality, character or purpose.

Infinite monkey theorem – Concept of **evolution** unrestrained by time, space or natural resources. Unthinkable in the real world but quite possible in the **cyberspace**.

Intranet – Network that's detached from the internet.

Large Hadron Collider – Largest scientific device ever made. Meant to find **Higgs boson** and reveal true nature of gravity and thus the universe.

Law of cause and effect – The idea that everything happens for a discernible and predictable reason. Powered scientific advancement for at least 2,000 years. Compare to **quantum entanglement**.

Learning – The process of associating observed information into universal rules that foster **evolution**. Slow and painstaking for humans. Compare to **machine learning**.

Limbic system – Brain core in humans and other animals. Instinctual and obsessed with survival. Controls breathing, heart rate, perspiration, etc.

Logic – A way to discover universal rules from easily observed events and properties.

Machine learning – Umbrella term for all processes meant to evolve computer programs through interaction with the real world or one another in a **black box** setting. Already in use on social media websites (i.e., Facebook's facial recognition **neural network**). Quaint name for **deep learning**.

Marshmallow test – 1960s experiment of leaving a child with a marshmallow for ten minutes to test their willpower. Supposedly measured self-restraint and later success in life. Called into doubt in 2018.

Mechanistic determination – A simplistic but workable idea of every state or action in nature being mathematically predictable. Compare to **quantum physics**.

Monte Carlo – **Algorithm** that searches through all available actions and their consequences to find the best one. Computationally burdensome and unfeasible in highly probabilistic games.

Neocortex – The outer layer of the human brain. Contains higher functions, such as personality traits and vision processing. Constantly struggles to control the **limbic system**.

Network jitter – Congestion of network packets that makes some of them late to the destination.

Neuralink – High-bandwidth brain implant to merge man and machine, creating a **cyborg**. Announced by Elon Musk. Unviable in the near future.

Neural network – Computer programs built to mimic the function, though not necessarily the form, of a living brain. Individual subroutines within the neural network serve the function of neurons in the brain. Trained using **machine learning**.

Neurotransmitter – Brain chemical, such as dopamine, that causes brain activity.

Paradox – A seemingly impossible conclusion that nonetheless appears to be gotten through **logic**. The cause is usually poor starting definition of terms.

Peer review – Academic practice of scientists screening new texts to determine the author's expertise before accepting his findings. Depends on there being enough reviewers willing to criticize.

Quantum entanglement – Notion that particles can have "soulmates"; affecting or just measuring one instantly affects the other over arbitrarily large distances and back through time. Einstein called it "spooky action at a distance". Currently unexplainable.

Quantum eraser experiment – Enhanced **double slit experiment**, with an added elaborate array of a crystal, mirrors, and detectors. Showed **quantum entanglement**. Implies the universe changes behavior when observed.

Quantum leap – Observable but unexplainable qualitative change, a major shift for the better that seemingly happened for no reason. Evolution from primate to man is one such example. Compare to **law of cause and effect**.

Quantum physics – A new scientific paradigm that states normal rules of physics don't apply on a molecular level. Drives classical physicists up the wall. Started by the **double slit experiment**.

Replication crisis – A long-standing problem of scientific studies having results or methods nobody can repeat. May turn scientists into a new priest class.

Robot – Mechanical worker. Originally from Czech robotnik, meaning "forced worker".

Schrödinger's cat – Thought experiment with a cat that's both alive and dead until someone observes it. An exception to the **Copenhagen interpretation** meant to disprove **quantum physics**. May cause a headache if pondered too much.

Scientific method – Gathering data to create a reproducible theory in a controlled setting. The basis of all civilization and technological progress.

Spectrogram – Representing time-variable signals on a 2D graph.

Stanford prison experiment – Psychological experiment from 1971. Supposedly showed humans are inherently psychopathic but nobody could replicate the results.

Steganalysis – Decrypting hidden data from public sources. Opposite of **steganography**.

Steganography – Weaving data into seemingly innocuous comments, images etc., that are released into the public for someone else to decode using **steganalysis**.

Super AI – Digital intelligence ascended to godlike levels. Expected to arise out of **general AI**. May become thoroughly unstable or change the course of evolution.

The butterfly effect – Idea that seemingly innocuous actions have long-term, global consequences. The term comes from a weather forecast model where a single rounded number completely changed the outcome two months down the line. Commonly misunderstood as "everything happens for a reason" but the true meaning is closer to "we can't know everything perfectly".

Tor – The Onion Router. A supposedly anonymous way to browse the internet using a network of relay nodes.

Tragedy of commons – Persistent, low-level destruction of the environment on a massive scale. Logical conclusion of a vast network of competing businesses sharing limited common resources and offloading their risk onto future generations.

TranslateGate – Discovery of ominous, bizarre and downright perverted messages when translating nonsensical inputs from Somali, Estonian, Maori etc., to English in Google Translate. Actual cause is that the neural network powering Google Translate isn't ready for real-time use, but Google can't resist padding its product portfolio with a translator service.

Unsupervised learning – Extremely fast process of a **neural network** associating different data points. It is unknown how unsupervised learning actually works, the focus is on the fact it does.

Voxel – Volumetric pixel aka "pixel in 3D". Mainly used in video games and graphic modeling up to this point.

www.ingramcontent.com/pod-product-compliance
Lightning Source LLC
Chambersburg PA
CBHW070847070326
40690CB00009B/1736